APPLIED COLLEGE ALGEBRA

By

A.K. Sharma

DISCOVERY PUBLISHING HOUSE PVT. LTD.
NEW DELHI-110 002

Published by:
Namit Wasan
DISCOVERY PUBLISHING HOUSE PVT. LTD.
4383/4B, Ansari Road, Darya Ganj
New Delhi-110 002 (India)
Phone : +91-11-23279245; 23253475; 43596065
E-mail : discoverybooksindia@gmail.com
discoverypublishinghouse@gmail.com
namitwasan9@gmail.com
web : www.discoverypublishinggroup.com

Edition: **2020**

ISBN: 978-81-8356-452-6

Applied College Algebra

Printed at:
Infinity Imaging Systems
Delhi

Preface

Algebra is one of the main branches of mathematics, concerning the study of structure, relation and quantity. Elementary algebra is often a part of the curriculum in secondary education and provides an introduction to the basic ideas in Algebra, including effects of adding and multiplying numbers, the concept of variables, polynomials, along with factorisation and determining their roots. Algebra is in fact much broader than elementary algebra and can be generalised. In addition to working directly, with numbers that algebra covers, while working with symbols, variables and set elements. Addition and multiplication are viewed as general operations, and their precise definitions lead to structures, such as groups, rings and fields.

Applied Algebra is the most useful form of Algebra. It is taught to students, who are presumed to have no basic knowledge of Algebra and are supposed to use Algebra, as a secondary subject in their study first, and profession later.

Although, there are many a good book on the subject, yet there is a demand for a precised book on Applied Algebra. So, we have made this humble effort to fill the same gap. Expectedly, this effort should prove to be beneficial for the students, teachers and general readers. Positive feedback is seriously welcome.

Content

Elementary Matrix Theory

When we introduce the various types of structures essential to the study of vectors and tensors, it is convenient in many cases to illustrate these structures by examples involving matrices. It is for this reason we are including a very brief introduction to matrix theory here. We shall not make any effort towards rigour in this chapter.

Real Matrix or Complex Matrix

An M by N matrix A is a rectangular array of real or complex numbers A_{ij} arranged in M rows and N columns. A matrix is often written

$$A = \begin{bmatrix} A_{11} & A_{12} & \cdot & \cdot & \cdot & A_{1N} \\ A_{21} & A_{22} & \cdot & \cdot & \cdot & A_{2N} \\ \cdot & & & & & \\ \cdot & & & & & \\ A_{M1} & A_{M2} & \cdot & \cdot & \cdot & A_{MN} \end{bmatrix} \tag{1.1}$$

and the numbers A_{ij} are called the *elements* or *components* of A. The matrix A is called a *real matrix* or a *complex matrix* according to whether the components of A are real numbers or complex numbers.

A matrix of M rows and N columns is said to be of *order* M by N or $M \times N$. It is customary to enclose the array with brackets, parentheses or double straight lines. We shall adopt the notation in (1.1). The location of the indices is sometimes modified to the forms A^{ij}, A^i_j, or A_i^j. Throughout this chapter the placement of the indices is unimportant and shall always be written as in (1.1). The elements $A_{i1}, A_{i2},..., A_{iN}$ are the elements of the ith row of A , and the elements $A_{1k}, A_{2k},..., A_{Nk}$ are the elements of the kth *column*. The convention is that the first index denotes the row and the second the column.

A row matrix is a $1\times N$ matrix, e.g.,

$$[A_{11}\ A_{12}\ .\ .\ .\ A_{1N}]$$

while a column matrix is an $M\times 1$ matrix, e.g.,

$$\begin{bmatrix} A_{11} \\ A_{21} \\ . \\ . \ . \\ . \\ A_{M1} \end{bmatrix}$$

The matrix A is often written simply

$$A = \left[A_{ij}\right] \tag{1.2}$$

Square Matrix and Identity Matrix

A *square matrix* is an $N\times N$ matrix. In a square matrix A, the elements $A_{11}, A_{22},...,A_{NN}$ are its are its diagonal elements. The sum of the diagonal elements of a square matrix A is called the *trace* and is written tr A. Two matrices A and B are said to be equal if they are identical. That is, A and B have the same number of rows and the same number of columns and

$$A_{ij} = B_{ij}, \qquad i = 1,...,N, \qquad j = 1,...,M$$

A matrix, every element of which is zero, is called the *zero matrix* and is written simply 0.

If $A=[A_{ij}]$ and $B = [B_{ij}]$ are two $M \times N$ matrices, their sum (difference) is an $M \times N$ matrix $A + B$ $(A - B)$ whose elements are $A_{ij} + B_{ij}$ $(A_{ij} - B_{ij})$. Thus

$$A \pm B = \left[A_{ij} \pm B_{ij}\right] \tag{1.3}$$

Two matrices of the same order are said to be conformable for addition and subtraction. Addition and subtraction are not defined for matrices which are not conformable. If λ is a number and A is a matrix, then λA is a matrix given by

$$\lambda A = \left[\lambda A_{ij}\right] = A\lambda \tag{1.4}$$

Therefore,

$$-A = (-1)A = \left[-A_{ij}\right] \tag{1.5}$$

These definitions of addition and subtraction and, multiplication by a number imply that

$$A + B = B + A \tag{1.6}$$

$$A + (B + C) = (A + B) + C \tag{1.7}$$

$$A + 0 = A \tag{1.8}$$

$$A - A = 0 \tag{1.9}$$

$$\lambda(A+B) = \lambda A + \lambda B \tag{1.10}$$

$$(\lambda+\mu)A = \lambda A + \mu A \tag{1.11}$$

and

$$1A = A \tag{1.12}$$

where A, B and C are as assumed to be conformable.

If A is an $M \times N$ matrix and B is an $N \times K$ matrix, then the product of B by A is written AB and is an $M \times K$ matrix with elements $\sum_{j=1}^{N} A_{ij} B_{js}$, $i = 1,\dots,M$, $s = 1,\dots,K$. For example, if

$$A = \begin{bmatrix} A_{11} & A_{12} \\ A_{21} & A_{22} \\ A_{31} & A_{32} \end{bmatrix} \text{ and } B = \begin{bmatrix} B_{11} & B_{12} \\ B_{21} & B_{22} \end{bmatrix}$$

then AB is a 3×2 matrix given by

$$AB = \begin{bmatrix} A_{11} & A_{12} \\ A_{21} & A_{22} \\ A_{31} & A_{32} \end{bmatrix} \begin{bmatrix} B_{11} & B_{12} \\ B_{21} & B_{22} \end{bmatrix}$$

$$= \begin{bmatrix} A_{11}B_{11} + A_{12}B_{21} & A_{11}B_{12} + A_{12}B_{22} \\ A_{21}B_{11} + A_{22}B_{21} & A_{21}B_{12} + A_{22}B_{22} \\ A_{31}B_{11} + A_{32}B_{21} & A_{32}B_{12} + A_{32}B_{22} \end{bmatrix}$$

The product AB is defined only when the number of columns of A is equal to the number of rows of B. If this is the case, A is said to be conformable to B for multiplication. If A is conformable to B, then B is not necessarily conformable to A. Even if BA is defined, it is not necessarily equal to AB. On the assumption that A, B, and C are conformable for the indicated sums and products, it is possible to show that

$$A(B + C) = AB + AC \tag{1.13}$$

$$(A + B)C = AC + BC \tag{1.14}$$

and

$$A(BC) = (AB)C \tag{1.15}$$

However, $AB \neq BA$ in general, $AB = 0$ does not imply $A = 0$ or $B = 0$, and $AB = AC$ does not necessarily imply $B = C$.

The square matrix I defined by

$$I = \begin{bmatrix} 1 & 0 & . & . & . & 0 \\ 0 & 1 & . & . & . & 0 \\ . & & & & & \\ . & & & & & \\ . & & & & & \\ 0 & 0 & . & . & . & 1 \end{bmatrix} \tag{1.16}$$

is the *identity matrix*. The identity matrix is a special case of a diagonal matrix which has all of its elements zero except the diagonal ones. A square matrix A whose elements satisfy $A_{ij} = 0$, $i > j$, is called *an upper triangular matrix,* i.e.,

$$A = \begin{bmatrix} A_{11} & A_{12} & A_{13} & . & . & . & A_{1N} \\ 0 & A_{22} & A_{23} & . & . & . & A_{2N} \\ 0 & 0 & A_{33} & . & . & . & \\ . & & & & & & \\ . & & & & & & \\ . & & & & & & \\ 0 & 0 & 0 & . & . & . & A_{NN} \end{bmatrix}$$

A lower triangular matrix can be defined in a similar fashion. A diagonal matrix is both an upper triangular matrix and a lower triangular matrix.

If A and B are square matrices of the same order such that $AB = BA = I$, then B is called the *inverse* of A and we write $B = A^{-1}$. Also, A is the inverse of B, i.e. $A = B^{-1}$. If A has an inverse it is said to be non-singular. If A and B are square matrices of the same order with inverses A^{-1} and B^{-1} respectively, then

$$(AB)^{-1} = B^{-1} A^{-1} \tag{1.17}$$

Equation (1.17) follows because

$$(AB)^{-1} B^{-1} A^{-1} = A(BB^{-1})A^{-1} = AIA^{-1} = AA^{-1} = I$$

and similarly

$$(B^{-1} A^{-1})AB = I$$

The matrix of order $N \times M$ obtained by interchanging the rows and columns of an $M \times N$ matrix A is called the *transpose* of A and is denoted by A^T. It is easily shown that

$$(A^T)^T = A \tag{1.18}$$

$$(\lambda A)^T = \lambda A^T \tag{1.19}$$

$$(A + B)^T = A^T + B^T \tag{1.20}$$

and

$$(AB)^T = B^T A^T \tag{1.21}$$

A square matrix A is *symmetric* if $A = A^T$ and *skew-symmetric* if $A = -A^T$. Therefore, for a summetric matrix

$$A_{ij} = A_{ji} \tag{1.22}$$

and for a skew-symmetric matrix

$$A_{ij} = -A_{ji} \tag{1.23}$$

Equation (1.23) implies that the diagonal elements of a skew symmetric-matrix are all zero. Every square matrix A can be written uniquely as the sum of a symmetric matrix and a skew-symmetric matrix, namely

$$A = \frac{1}{2}(A+A^T)+\frac{1}{2}(A-A^T) \tag{1.24}$$

If A is a square matrix, its determinant is written det A. The reader is assumed at this stage to have some familiarity with the computational properties of determinants. In particular, it should be known that

$$\det A = \det A^T \text{ and } \det AB = (\det A)(\det B) \tag{1.25}$$

If A is an $N \times N$ square matrix, we can construct an $(N - 1) \times (N - 1)$ square matrix by removing the ith row and the jth column. The determinant of this matrix is denoted by M_{ij} and is the *minor* of A_{ij}. The *cofactor* of A_{ij} is defined by

$$\text{cof } A_{ij} = (-1)^{i+j} M_{ij} \tag{1.26}$$

For example, if

$$A = \begin{bmatrix} B_{11} & B_{12} \\ B_{21} & B_{22} \end{bmatrix} \tag{1.27}$$

then

$$M_{11} = A_{22}, M_{12} = A_{21} \quad M_{21} = A_{12} \quad M_{22} = A_{11}$$

$$\text{cof} = A_{11} = A_{22} \text{ cof } A_{12} = -A_{21}, \text{ etc.}$$

The adjoint of an $N \times N$ matrix A, written adj A, is an $N \times N$ matrix given by

$$\text{adj}\,A = \begin{bmatrix} \text{cof}\,A_{11} & \text{cof}\,A_{21} & . \;\; . \;\; . & \text{cof}\,A_{N1} \\ \text{cof}\,A_{12} & \text{cof}\,A_{22} & . \;\; . \;\; . & \text{cof}\,A_{N2} \\ . & & & \\ . & & & \\ . & & & \\ \text{cof}\,A_{1N} & \text{cof}\,A_{2N} & . \;\; . \;\; . & \text{cof}\,A_{NN} \end{bmatrix} \tag{1.28}$$

The reader is cautioned that the designation "adjoint" is used in a different context. It is possible to show that

$$A\ (\text{adj } A) = (\det A)\ I = (\text{adj } A)\ A \tag{1.29}$$

We shall prove (1.29) in general, later; so we shall be content to establish it for $N = 2$ here. For $N = 2$

$$\text{adj}\,A = \begin{bmatrix} A_{22} & -A_{12} \\ -A_{21} & A_{11} \end{bmatrix}$$

Then

$$A(\text{adj}\,A) = \begin{bmatrix} A_{11} & A_{12} \\ A_{21} & A_{22} \end{bmatrix} \begin{bmatrix} A_{22} & -A_{21} \\ -A_{21} & A_{11} \end{bmatrix}$$

$$= \begin{bmatrix} A_{11}A_{22} - A_{12}A_{21} & 0 \\ 0 & A_{11}A_{22} - A_{12}A_{21} \end{bmatrix}$$

Therefore

$$A\ (\text{adj } A) = (A_{11}A_{22} - A_{12}A_{21})\ I = (\det A)\ I$$

Likewise

$$(\text{adj } A)\ A = (\det A)\ I$$

If $\det A \neq 0$, then (1.29) shows that the inverse A^{-1} exists and is given by

$$A^{-1} = \frac{\text{adj}\,A}{\det A} \tag{1.30}$$

Of course, if A^{-1} exists, then $(\det A)\ (\det A^{-1}) = \det I = 1 \neq 0$. Thus non-singular matrices are those with a non-zero determinant.

Matrix Notation

Matrix notation is convenient for manipulations of systems of linear algebraic equations. For example, if we have the set of equations

$$A_{11}x_1 + A_{12}x_2 + A_{13}x_3 + \cdots + A_{1N}x_N = Y_1$$
$$A_{21}x_1 + A_{22}x_2 + A_{23}x_3 + \cdots + A_{2N}x_N = Y_2$$
$$\vdots$$
$$A_{N1}x_1 + A_{N2}x_2 + A_{N3}x_3 + \cdots + A_{NN}x_N = Y_2$$

then they can be written

$$\begin{bmatrix} A_{11} & A_{12} & \cdots & A_{1N} \\ A_{21} & A_{22} & & A_{2N} \\ \cdot & & & \\ \cdot & & & \\ \cdot & & & \\ A_{N1} & A_{N2} & \cdots & A_{NN} \end{bmatrix} \begin{bmatrix} x_1 \\ x_2 \\ \cdot \\ \cdot \\ \cdot \\ x_N \end{bmatrix} = \begin{bmatrix} y_1 \\ y_2 \\ \cdot \\ \cdot \\ \cdot \\ y_N \end{bmatrix}$$

The above matrix equation can now be written in the compact notation

$$AX = Y \tag{1.31}$$

and if A is non-singular the solution is

$$X = A^{-1}Y \tag{1.32}$$

For $N = 2$, we can immediately write

$$\begin{bmatrix} x_1 \\ x_2 \end{bmatrix} = \frac{1}{\det A} \begin{bmatrix} A_{22} & -A_{12} \\ -A_{21} & A_{11} \end{bmatrix} \begin{bmatrix} y_1 \\ y_2 \end{bmatrix} \tag{1.33}$$

Exercises

1. Add the matrices

$$\begin{bmatrix} 1 \\ 2 \end{bmatrix} + \begin{bmatrix} -5 \\ 6 \end{bmatrix}$$

Add the matrices

$$\begin{bmatrix} 2i & 3 & 7+2i \\ 5 & 4+3i & i \end{bmatrix} + \begin{bmatrix} 2 & 5i & -4 \\ -4i & 3+3i & -i \end{bmatrix}$$

2. Add

$$\begin{bmatrix} 1 \\ 2i \end{bmatrix} + 3\begin{bmatrix} -5i \\ 6 \end{bmatrix}$$

3. Multiply

$$\begin{bmatrix} 2i & 3 & 7+2i \\ 5 & 4+3i & i \end{bmatrix}\begin{bmatrix} 2i & 8 \\ 1 & 6i \\ 3i & 2 \end{bmatrix}$$

4. Show that the product of two upper (lower) triangular matrices is an upper lower triangular matrix. Further, if

$$A = [A_{ij}], \quad B = [B_{ij}]$$

are upper (lower) triangular matrices of order $N \times N$, then

$$(AB)_{ii} = (BA)_{ii} = A_{ii}B_{ii}$$

for all $i = 1,\ldots,N$. The off diagonal elements $(AB)_{ij}$ and $(BA)_{ij}$, $i \neq j$, generally are not equal, however.

5. What is the transpose of

$$\begin{bmatrix} 2i & 3 & 7+2i \\ 5 & 4+3i & i \end{bmatrix}$$

6. What is the inverse of

$$\begin{bmatrix} 5 & 3 & 1 \\ 0 & 7 & 2 \\ 1 & 4 & 0 \end{bmatrix}$$

7. Solve the equations

$$5x + 3y = 2$$
$$x + 4y = 9$$

by use of (1.33).

Sets, Relations and Functions

The purpose of this chapter is to introduce an initial vocabulary and some basic concepts. Most of the readers are expected to be somewhat familiar with this material; so its treatment is brief.

Sets and Set Algebra

The concept of a set is regarded here as primitive. It is a collection or family of things viewed as a simple entity. The things in the set are called *elements* or *members* of the set. They are said to be contained in or to belong to the set. Sets will generally be denoted by upper case script letters, $A, B, C, D, \ldots$, and elements by lower case letters, $a, b, c, d, \ldots$. The sets of complex, numbers, real numbers and integers will be denoted by C, R, and I, respectively. The notation $a \in A$ means that the element a is contained in the set A; if a is not an element of A, the notation $a \notin A$ is employed.

To denote the set whose elements are a, b, c, and d the notation $\{a, b, c, d\}$ is employed. In mathematics, a set is not generally a collection of unrelated objects like a tree, a doorknob and a concept, but rather it is a collection which share some common property like the vineyards of France which share the common property of being in France or the real numbers which share the common property of being real. A set whose elements are determined by their possession of a certain property is denoted by $\{x \mid P(x)\}$, where x denotes a typical element and $P(x)$ is the property which determines x to be in the set.

If A and B are sets, B is said to be a subset of A if every element of B is also an element of A. It is customary to indicate that B is a subset of A by the notation $B \subset A$, which may be read as "B is contained in A," or $A \supset B$ which may be read as "A contains B." For example, the set of integers I is a subset of the set of real numbers R, $I \subset R$. Equality of two sets A and B is said to exist if A is a subset of B and B is a subset of A; in equation form

$$A = B \Leftrightarrow A \subset B \text{ and } B \subset A \tag{2.1}$$

A non-empty subset B of A is called a *proper subset* of A if B is not equal to A. The set of integers I is actually a proper subset of the real numbers. The *empty set* or *null set* is the set with no elements and is denoted by $\varnothing$. The *singleton* is the set containing a single element a and is denoted by $\{a\}$. A set whose elements are sets is often called a *class*.

Some operations on sets which yield other sets will now be introduced. The *union* of the sets A and B is the set of all elements that are either in the set A or in the set B. The union of A and B is denoted by $A \cup B$

$$A \cup \mathrm{B} \equiv \{a \mid a \in A \text{ or } a \in B\} \tag{2.2}$$

It is easy to show that the operation of forming a union is commutative,

$$A \cup B = B \cup \mathrm{A} \tag{2.3}$$

and associative

$$A \cup (B \cup \mathrm{C}) = (A \cup B) \cup \mathrm{C} \tag{2.4}$$

The intersection of the sets A and B is the set of elements that are in both A and B. The intersection is denoted by $A \cap B$ and is specified by

$$A \cap \mathrm{B} = \{a \mid a \in A \text{ and } a \in B\} \tag{2.5}$$

Again, it can be shown that the operation of intersection is commutative

$$A \cap B = B \cap A \tag{2.6}$$

and associative

$$A \cap (B \cap C) = (A \cap B) \cap C \tag{2.7}$$

Two sets are said to be disjoint if they have no elements in common, i.e. if

$$A \cap B = \varnothing \tag{2.8}$$

The operations of union and intersection are related by the following distributive laws:

$$\begin{aligned} A \cap (B \cup C) &= (A \cap B) \cup (A \cap C) \\ A \cup (B \cap C) &= (A \cup B) \cap (A \cup C) \end{aligned} \tag{2.9}$$

The complement of the set B with respect to the set A is the set of all elements contained in A but not contained in B. The complement of B with respect to the set A is denoted by $A | B$ and is specified by

$$A | B = \{a \,|\, a \in A \text{ and } a \notin B\} \tag{2.10}$$

It is easy to show that

$$A | (A | B) = B \quad \text{for} \quad B \subset A \tag{2.11}$$

and

$$A | B = A \Leftrightarrow A \cap B = \varnothing \tag{2.12}$$

Exercises

1. List all possible subsets of the set $\{a, b, c, d\}$.
2. List two proper subsets of the set $\{a, b,\}$.
3. Prove the following formulas:
   (a) $A = A \cup B \Leftrightarrow B \subset A$.
   (b) $A = A \cup \varnothing$.
   (c) $A = A \cup A$.

4. Show that $I \cap R = I$.
5. Verify the distributive laws (2.9).
6. Verify the following formulas:
   a. $A \subset B \Leftrightarrow A \cap B = A$.
   b. $A \cap \varnothing = \varnothing$.
   c. $A \cap A = A$.
7. Give a proof of the commutative and associative properties of the intersection operation.
8. Let $A = \{-1,-2,-3,-4\}$, $B = \{-1,0,1,2,3,7\}$, $C = \{0\}$, and $D = \{-7,-5,-3,-1,1,2,3\}$. List the elements of $A \cup B$, $A \cap B$, $A \cup C$, $B \cap C$, $A \cup B, \cup C$, $A \mid B$, and $(D \mid C) \cup A$.

Ordered Pairs, Cartesian Products and Relations

The idea of ordering is not involved in the definition of a set. For example, the set $\{a,b\}$ is equal to the $\{b,a\}$. In many cases of interest it is important to order the elements of a set. To define an *ordered pair* (a,b) we single out a particular element of $\{a,b\}$, say a, and define (a,b) to be the class of sets; in equation form,

$$(a,b) \equiv \{\{b\},\{a,b\}\} \tag{2.13}$$

This ordered pair is then different from the ordered pair (b,a) which is defined by

$$(b,a) \equiv \{\{b\},\{b,a\}\} \tag{2.14}$$

It follows directly from the definition that

$$(a,b) = (c,d) \Leftrightarrow a = c \text{ and } b = d$$

Clearly, the definition of an ordered pair can be extended to an ordered N -tuple $(a_1,a_2,\ldots,a_N)$.

The Cartesian Product of two sets A and B is a set whose elements are ordered pairs (a,b) where a is an element of A and b is an element of B. We denote the Cartesian product of A and B by $A \times B$.

$$A \times B = \{(a,b) \mid a \in A,\ b \in B\} \tag{2.15}$$

It is easy to see how a Cartesian product of N sets can be formed using the notion of an ordered N- tuple.

Any subset K of $A \times B$ defines a relation from the set A to the set B. The notation aKb is employed if $(a,b) \in K$; this notation is modified for negation as aKb if $(a,b) \notin K$. As an example of a relation, let A be the set of all Volkswagons in the United States and let B be the set of all Volkswagons in Germany. Let be $K \subset A \times B$ be defined so that aKb if b is the same colour as a.

If $A = B$, then K is a relation on A. Such a relation is said to be *reflexive* if a aKa for all $a \in A$, it is said to be *symmetric* if aKb whenever bKa for any a and b in A, and it is said to be transitive if aKb and bKc imply that aKc for any a,b and c in A. A relation on a set A is said to be an *equivalence relation* if it is reflexive, symmetric and transitive. As an example of an equivalence relation K on the set of all real numbers R let $aKb \Leftrightarrow a = b$ for all a,b. To verify that this is an equivalence relation, note that $a = a$ for each $a \in K$ (reflexivity), $a = b \Rightarrow b = a$ for each $a \in K$ (symmetry), and $a = b$, $b = c \Rightarrow a =$c for a,b and c in K (transitivity).

A relation K on A is antisymmetric if aKb then bKa unless $b = a$. A relation on a set A is said to be a *partial ordering* if it is reflexive, anti-symmetric and transitive. The equality relation is a trivial example of partial ordering. As another example of a partial ordering on R, let the inequality $\leq$ be the relation. To verify that this is a partial ordering note that $a \leq a$ for every $a \in R$ (reflexivity), $a \leq b$ and $b \leq a \Rightarrow a = b$ for all $a,b \in R$ (anti-symmetry), and $a \leq b$ and $b \leq c \Rightarrow a \leq c$ for all $a,b,c \in R$ (transitivity). Of course, inequality is not an equivalence relation since it is not symmetric.

Exercises

1. Define an ordered triple (a,b,c) by

   (a,b,c) = $((a,b),c)$

   and show that $(a,b,c) = (d,e,f) \Leftrightarrow a = d$, $b = e$, and $c = f$.

2. Let A be a set and K an equivalence relation on A. For each $a \in A$ consider the set of all elements x that stand in the

relation K to a; this set is denoted by $\{x \mid aKx\}$, and is called an *equivalence set.*

(a) Show that the equivalence set of a contains a.

(b) Show that any two equivalence sets either coincide or are disjoint.

Note: (a) and (b) show that the equivalence sets form a partition of A; A is the disjoint union of the equivalence sets.

3. On the set R of all real numbers does the strict inequality $<$ constitute a partial ordering on the set?

4. For ordered triples of real numbers define $(a_1, b_1, c_1) \leq (a_2, b_2, c_2)$ if $a_1, \leq a_2, b_1 \leq b_2, c_1 \leq c_2$. Does this relation define a partial ordering on $R \times R \times R$?

Various Functions

A relation f from X to Y is said to be a *function* (or a mapping) if $(x, y_1) \in f$ and $(x, y_2) \in f$ imply $y_1 = Y_2$ for all $x \in X$. A function is thus seen to be a particular kind of relation which has the property that for each $x \in X$ there exists one and only one $y \in Y$ such that $(x,y) \in f$. Thus a function f defines a *single-valued relation.* Since a function is just a particular relation, the notation of the last section could be used, but this is rarely done. A standard notation for a function from X to Y is

$$f: X \rightarrow Y$$

or

$$y = f(x)$$

which indicates the element $y \in Y$ that f associates with $x \in X$. The element y is called the *value* of f at x. The domain of the function f is the set X. The *range* of f is the set of all y for which there exists an x such that $y = f(x)$. We denote the range of f by $f(X)$. When the domain of f is the set of real numbers R, f is said to be a *function of a real variable.* When $f(X)$ is contained in R, the function is said to be *real-valued.* Functions of a complex variable and complex-valued functions are defined analogously. A function of a real variable need not be real-valued, nor need a function of a complex variable be complex-valued.

If X_0 is a subset of $X, X_0 \subset X$, and $f: X \rightarrow Y$, the image of X_0 under f is the set

$$f(X_0) \equiv \{f(x) \mid x \in X_0\} \tag{2.16}$$

and it is easy to prove that

$$f(X_0) \subset f(X) \tag{2.17}$$

Similarly, if Y_0 is a subset of Y, then the preimage of Y_0 under f is the set

$$f^{-1}(Y_0) \equiv \{x \mid f(x) \in Y_0\} \tag{2.18}$$

and it is easy to prove that

$$f^{-1}(Y_0) \subset X \tag{2.19}$$

A function f is said be into Y if $f(X)$ is a proper subset of Y, and it is said to be *onto* Y if $f(X) = Y$. A function that is onto is also called *surjective*. Stated in a slightly different way, a function is onto if for every element $y \in Y$ there exists at least one element $x \in X$ such that $y = f(x)$. A function f is said to be *one-to-one* (or *injective*) if for every $x_1, x_2 \in X$

$$f(x_1) = f(x_2) \Rightarrow x_1 = x_2 \tag{2.20}$$

In other words, a function is one-to-one if for every element $y \in f(X)$ there exists only one element $x \in X$ such that $y = f(x)$. A function f is said to form a *one-to-one correspondence* (or to be *bijective*) from X to Y if $f(X) = Y$, and f is one-to-one. Naturally, a function can be onto without being one-to-one or be one-to-one without being onto. The following examples illustrate these terminologies.

1. Let $f(x) = x^5$ be a particular real valued function of a real variable. This function is one-to-one because $x^5 = z^5$ implies $x = z$ when $x, z \in R$ and it is onto since for every $y \in R$ there is some x such that $y = x^5$.
2. Let $f(x) = x^3$ be a particular complex-valued function of a complex variable. Thisfunction is onto but it is not one-to-one because, for example,

$$f(1) = f(-\frac{1}{2} + i\frac{\sqrt{3}}{2}) = f(-\frac{1}{2} - i\frac{\sqrt{3}}{2}) = 1^3$$

where $i = \sqrt{-1}$.

3. Let $f(x) = 2|x|$ be a particular real valued function of a real variable where $|x|$ denotes the absolute value of x. This function is into rather than onto and it is not one-to-one.

If $f: X \to Y$ is a one-to-one function, then there exists a function $f^{-1}: f(X) \to X$ which associates with each $y \in f(X)$ the unique $x \in X$ such that $y = f(x)$. We call f^{-1} the *inverse* of f, written $x = f^{-1}(y)$. Unlike the preimage defined by (2.18), the definition of inverse functions is possible only for functions that are one-to-one. Clearly, we have $f^{-1}(f(X))$. The *composition* of two functions $f: X \to Y$ and $g: \to Y \to Z$ is a function $h: X \to Z$ defined by $h(x) = g(f(x))$ for all $x \in X$. The function h is written $h = g \circ f$. The composition of any finite number of functions is easily defined in a fashion similar to that of a pair. The operation of composition of functions is not generally commutative,

$$g \circ f \neq f \circ g$$

Indeed, if $g \circ f$ is defined, $f \circ g$ may not be defined, and even if $g \circ f$ and $f \circ g$ are both defined, they may not be equal. The operation of composition is associative

$$h \circ (g \circ f) = (h \circ g) \circ f$$

if each of the indicated compositions is defined. The identity function $id: X \to X$ is defined as the function $id(x) = x$ for all $x \in X$. Clearly if f is a one-to-one correspondence from X to Y, then

$$f^{-1} \circ f = id_X, \qquad f \circ f^{-1} = id_y$$

where id_X and id_Y denote the identity functions of X and Y, respectively.

To close this discussion of functions we consider the special types of functions called sequences. A finite sequence of N terms is a function f whose domain is the set of the first N positive integers $\{1,2,3,..., N\}$. The range of f is the set of N elements $\{f(1), f(2),...,f(N)\}$, which is usually denoted by $\{f_1, f_2,...,f_N\}$. The elements

$f_1, f_2,...,f_N$ of the range are called *terms* of the sequence. Similarly, an *infinite sequence* is a function defined on the positive integers I^+. The range of an infinite sequence is usually denoted by $\{f_n\}$, which stands for the infinite set $\{f_1, f_2, f_3,...\}$. A function g whose domain is I^+ and whose range is contained in I^+ is said to be *order preserving* if $m < n$ implies that $g(m) < g(n)$ for all $m,n \in I^+$. If f is an infinite sequence and g is order preserving, then the composition $f \circ g$ is a subsequence of f. For example, let

$$f_n = 1/(n+1), \qquad g_n = 4$$

Then

$$f \circ g(n) = 1/(4^n + 1)$$

is a subsequence of f.

Exercises

1. Verify the results (2.17) and (2.19).
2. How may the domain and range of the sine function be chosen so that it is a one-to-one correspondence?
3. Which of the following conditions defines a function $y = f(x)$?

   $$x^2 + y^2 = 1, \qquad xy = 1, \qquad x,y \in R.$$
4. Give an example of a real valued function of a complex variable.
5. Can the implication of (2.20) be reversed? Why?
6. Under what conditions is the operation of composition of functions commutative?
7. Show that the arc sine function $\sin^{-1}$ really is not a function from R to R. How can it be made into a function?

Groups, Rings and Fields

The mathematical concept of a group has been crystallised and abstracted from numerous situations both within the body of mathematics and without. Permutations of objects in a set is a group operation.

The operations of multiplication and addition in the real number system are group operations. The study of groups is developed from the study of particular situations in which groups appear naturally, and it is instructive that the group itself be presented as an object for study.

Axioms for a Group

Central to the definition of a group is the idea of a binary operation. If G is a non-empty set, a binary operation on G is a function from $C \times C$ to G. If a, b $\in G$, the binary operation will be denoted by $*$ and its value by $a*b$. The important point to be understood about a binary operation on G is that G is closed with respect to $*$ in the sense that if a, b $\in G$ then $a *$ b $\in G$ also. A binary operation $*$ on G is associative if

$$(a^*b)^*c = a\ ^*(b^*c) \text{ for all } a,b,c, \in G \tag{3.1}$$

Thus, parentheses are unnecessary in the combination of more than two elements by an associative operation. A semi-group is a pair $(G,*)$ consisting of a non-empty set G with an associative binary operation$*$. A binary operation $*$ on G is commutative if

$$a*b = b*a \quad \text{for all } a, b, \in G \tag{3.2}$$

The multiplication of two real numbers and the addition of two real numbers are both examples of commutative binary operations. The multiplication of two N by N matrices ($N > 1$) is a non-commutative binary operation. A binary operation is often denoted by $a \circ b$, a · b, or ab rather than by $a*b$; if the operation is commutative, the notation $a + b$ is sometimes used in place of $a*b$.

An element $e \in G$ that satisfies the condition

$$e*a = a*e = a \quad \text{for all } a \in G \tag{3.3}$$

is called an identity element for the binary operation * on the set G. In the set of real numbers with the binary operation of multiplication, it is easy to see that the number 1 plays the role of identity element. In the set of real numbers with the binary operation of addition, 0 has the role of the identity element. Clearly G contains at most one identity element in e. For if e^1 is another identity element in G, then $e^1 *a = a * e^1 =$ a for all $a \in G$ also. In particular, if we choose $a = e$, then $e^1 *e =$ e. But from (3.3), we have also$e^1 *e = e^1$. Thus $e^1 =$ e. In general, G need not have any identity element. But if there is an identity element, and if the binary operation is regarded as multiplicative, then the identity element is often called the unity element; on the other hand, if the binary operation is additive, then the identity element is called the *zero element*.

In a semi-group G containing an identity element e with respect to the binary operation $*$, an element a^{-1} is said to be an inverse of the element a if

$$a*a^{-1} = a^{-1} * a = e \tag{3.4}$$

In general, a need not have an inverse. But if an inverse a^{-1} of a exists, then it is unique, the proof being essentially the same as that of the uniqueness of the identity element. The identity element is its own inverse. In the set $R/\{0\}$ with the binary operation of multiplication, the inverse of a number is the reciprocal of the number. In the set of real numbers with the binary operation of addition the inverse of a number is the negative of the number.

A group is a pair $(G,*)$ consisting of an associative binary operation $*$ and a set G which contains the identity element and the inverses of all elements of G with respect to the binary operation$*$. This definition can be explicitly stated in equation form as follows.

Definition: A group is a pair $(G,*)$ where G is a set and $*$ is a binary operation satisfying the following:

(a) $(a*b)*c = a*(b*c)$ for all $a,b,c \in G$.

(b) There exists an element $e \in G$ such that $a*e = e*a = a$ for all $a \in G$.

(c) For every $a \in G$ there exists an element $a^{-1} \in G$ such that $a*a^{-1} = a^{-1}*a = e$.

If the binary operation of the group is commutative, the group is said to be a *commutative* (or *Abelian*) *group*.

The set $R/\{0\}$ with the binary operation of multiplication forms a group, and the set R with the binary operation of addition forms another group. The set of positive integers with the binary operation of multiplication forms a semi-group with an identity element but does not form a group because the condition (c) above is not satisfied.

A notational convention customarily employed is to denote a group simply by G rather than by the pair $(G,*)$. This convention assumes that the particular $*$ to be employed is understood. We shall follow this convention here.

Exercises

1. Verify that the set $G \in \{1,i,-1,-i\}$, where $i^2 = -1$, is a group with respect to the binary operation of multiplication.
2. Verify that the set G consisting of the four 2×2 matrices

$$\begin{bmatrix} 1 & 0 \\ 0 & 1 \end{bmatrix}, \quad \begin{bmatrix} 0 & 1 \\ -1 & 0 \end{bmatrix}, \quad \begin{bmatrix} 0 & -1 \\ 1 & 0 \end{bmatrix}, \quad \begin{bmatrix} -1 & 0 \\ 0 & -1 \end{bmatrix}$$

constitutes a group with respect to the binary operation of matrix multiplication.

3. Verify that the set G consisting of the four 2×2 matrices

$$\begin{bmatrix} 1 & 0 \\ 0 & 1 \end{bmatrix}, \quad \begin{bmatrix} -1 & 0 \\ 0 & 1 \end{bmatrix}, \quad \begin{bmatrix} 1 & 0 \\ 0 & -1 \end{bmatrix}, \quad \begin{bmatrix} -1 & 0 \\ 0 & -1 \end{bmatrix}$$

constitutes a group with respect to the binary operation of matrix multiplication.

4. Determine a subset of the set of all 3×3 matrices with the binary operation of matrix multiplication that will form a group.

5. If A is a set, show that the set G, $G = \{f \mid f: A \rightarrow A, f$ is a one-to-one correspondence$\}$ is a one-to-one correspondence$\}$ constitutes a group with respect to the binary operation of composition.

Properties of a Group

In this section, certain basic properties of groups are established. There is a great economy of effort in establishing these properties for a group in general because the properties will then be possessed by any system that can be shown to constitute a group, and there are many such systems of interest in this text. For example, any property established in this section will automatically hold for the group consisting of $R/\{0\}$ with the binary operation of multiplication, for the group 0 / consisting of the real numbers with the binary operation of addition, and for groups involving vectors and tensors.

The basic properties of a group G in general are:

1. The identity element $e \in G$ is unique.

2. The inverse element a^{-1} of any element $a \in G$ is unique. The proof of Property 1 has been given. The proof of Property 2 follows by essentially the same argument.

   If n is a positive integer, the powers of $a \in G$ are defined as follows: (i) For $n = 1$, $a^1 = a$. (ii) For $n > 1$, $a^{n-1} * a$. (iii) $a^0 = e$. (iv) $a^{-n} = (a^{-1})^n$.

3. If m,n,k are any integers, positive or negative, then for a $\in G$

$$a^m * a^n = a^{m+n},\ (a^m)^n = a^{mn},\ (a^{m+n})^k = a^{mk+nk}$$

In particular, when $m = n = -1$, we have:

4. $(a^{-1})^{-1} = a$ for all $a \in G$
5. $(a * b)^{-1} = b^{-1} * a^{-1}$ for all $a,b \in G$,

   For square matrices the proof of this property is given by (1.17); the proof in general is exactly the same.

   The following property is a useful algebraic rule for all groups; it gives the solutions x and y to the equations $x * a = b$ and $a * y = b$, respectively.

6. For any elements a,b in G, the two equations $x * a = b$ and $a * y = b$ have the unique solutions $x = b * a^{-1}$ and $y = a^{-1} * b$.

Proof: Since the proof is the same for both equations, we will consider only the equation $x * a = b$. Clearly $x = b * a^{-1}$ satisfies the equation $x * a = b$,

$$x * a - (b * a^{-1}) * a = b * (a^{-1} * a) = b * e = b$$

Conversely, $x * a = b$, implies

$$x = x * e = x * (a * a^{-1}) = (x * a) * a^{-1} = b * a^{-1}$$

which is unique. As a result we have the next two properties.

7. For any three elements a,b,c in G, either $a * c = b * c$ or $c * a = c * b$ implies that $a = b$.
8. For any two elements a,b in the group G, either $a * b = b$ or $b * a = b$ implies that a is the identity element.

   A non-empty subset G' of G is a subgroup of G if G' is a group with respect to the binary operation of G, i.e., G' is a subgroup of G, if and only if: (*i*) $e \in G$, (ii) $a \in G' \Rightarrow a^{-1} \in G'$ (iii) $a,b, \in G' \Rightarrow a * b \in G'$.

9. Let G' be a non-empty subset of G. Then G' is a subgroup if $a,b \in G' \Rightarrow G'$.

Proof: The proof consists in showing that the conditions (i), (ii), (iii) of a subgroup are satisfied. (i) Since G' is non-empty, it contains an element a, hence $a * a^{-1} = e \in G$. (ii) If $b \in G'$ then $e * b^{-1} = b^{-1} \in G'$. (iii) If $a,b \in G'$, then $a * (b^{-1})^{-1} = a * b \in G'$.

If G is a group, then G itself is a subgroup of G, and the group consisting only of the element e is also a subgroup of G. A subgroup of G other than G itself and the group e is called a *proper subgroup* of G.

10. The intersection of any two subgroups of a group remains a subgroup.

Exercises

1. Show that e is its own inverse.
2. Prove Property 3.
3. Prove Properties 7 and 8.
4. Show that $x = y$ in Property 6 if G is Abelian.
5. If G is a group and $a \in G$ show that $a * a = a$ implies $a = e$.
6. Prove Property 10.
7. Show that if we look at the non-zero real numbers under multiplication, then: (a) the rational numbers form a subgroup, (b) the positive real numbers form a subgroup, (c) the irrational numbers do not form a subgroup.

Group Homomorphisms

A "group homomorphism" is a fairly overpowering phrase for those who have not heard it or seen it before. The word homomorphism means simply the same type of formation or structure. A group homomorphism has then to do with groups having the same type of formation or structure.

Specifically, if G and H are two groups with the binary operations $*$ and $\circ$, respectively, a function $f : G \to H$ is a *homomorphism* if

$$f(a * b) = f(a) \circ f(b) \quad \text{for all } a,b \in G \qquad (3.5)$$

If a homomorphism exists between two groups, the groups are said to be homomorphic. A homomorphism $f : G \to H$ is an *isomorphism* if f is both one-to-one and onto. If an isomorphism exists between two groups, the groups are said to be *isomorphic*. Finally, a homomorphism $f: G \to G$ is called an *endomorphism* and an isomorphism $f: G \to G$ is called an *automorphism*. To illustrate these definitions, consider the following examples:

1. Let G be the group of all non-singular, real, $N \times N$ matrices with the binary operation of matrix multiplication. Let H be the group $R/\{0\}$ with the binary operation of scalar multiplication. The function that is the determinant of a matrix is then a homomorphism from G to H.
2. Let G be any group and let H be the group whose only element is e. Let f be the function that assigns to every element of G the value e; then f is a (trivial) homomorphism.
3. The identity function id: $G \to G$ is a (trivial) automorphism of any group.
4. Let G be the group of positive real numbers with the binary operation of multiplication and let H be the group of real numbers with the binary operation of addition. The log function is an isomorphism between G and H.
5. Let G be the group $R/\{0\}$ with the binary operation of multiplication. The function that takes the absolute value of a number is then an endomorphism of G into G. The restriction of this function to the subgroup H of H consisting of all positive real numbers is a (trivial) automorphism of H, however.

From these examples of homomorphisms and automorphisms one might note that a homomorphism maps identities into identities and inverses into inverses. The proof of this observation is the content of Theorem 3.1 below. Theorem 3.2 shows that a homomorphism takes a subgroup into a subgroup and Theorem 3.3 proves a converse result.

Theorem 3.1: If $f: G \to H$ is a homomorphism, then $f(e)$ coincides with the identity element e_0 of H and

$$f(a^{-1}) = f(a)^{-1}$$

Proof: Using the definition (3.1) of a homomorphism on the identity $a = a * e$, one finds that

$$f(a) = f(a * e) = f(a) \circ f(e)$$

From Property 8 of a group it follows that $f(e)$ is the identity element e_0 of H. Now, using this result and the definition of a homomorphism (3.1) applied to the equation $e = a * a^{-1}$, we find

$$e_0 = f(e) = f(a * a^{-1}) = f(a) \circ f(a^{-1})$$

It follows from Property 2 that $f(a^{-1})$ is the inverse of $f(a)$.

Theorem 3.2: If f: $G \rightarrow H$ is a homomorphism and if G' is a subgroup of G, then $f(G')$ is a subgroup of H.

Proof: The proof will consist in showing that the set $f(G')$ satisfies the conditions (i), (ii), (iii) of a subgroup. (i) Since $e \in G'$ it follows from Theorem 3.1 that the identity element e_0 of H is contained in $f(G')$. (ii) For any $a \in G'$, $f(a)^{-1} \in f(G')$ by Theorem 3.1. (iii) For any $a,b \in G'$ $f(a) \circ f(b) \in f(G')$ since $f(a) \circ f(b) = f(a*b) \in f(G')$.

As a corollary to Theorem 3.2 we see that $f(G)$ is itself a subgroup of H.

Theorem 3.3: If f: $G \rightarrow H$ is a homomorphism and if H' is a subgroup of H, then the preimage $f^{-1}(H')$ is a subgroup of G.

The kernel of a homomorphism f: $G \rightarrow H$ is the subgroup $f^{-1}(e_0)$ of G. In other words, the kernel of f is the set of elements of G that are mapped by f to the identity element e_0 of H.

The notation $K(f)$ will be used to denote the kernel of f. In Example 1 above, the kernel of f consists of $N \times N$ matrices with determinant equal to the real number 1, while in Example 2 it is the entire group G. In Example 3 the kernel of the identity map is the identity element, of course.

Theorem 3.4: A homomorphism f: $G \rightarrow H$ is one-to-one if and only if $K(f) = \{e\}$.

Proof: This proof consists in showing that $f(a) = f(b)$ implies $a = b$ if and only if $K(f) = \{e\}$. If $f(a) = f(b)$, then $f(a) \circ f(b)^{-1} = e_0$, hence $f(a * b^{-1}) = e_0$, and it follows that $a * b^{-1} \in K(f)$. Thus, if $K(f) = \{e\}$, then $a * b^{-1} = e$ or $a = b$. Conversely, now, we assume f is one-to-one and since $K(f)$ is a subgroup of G by Theorem 3.3, it must contain e. If $K(f)$ contains any other element a such that $f(a) = e_0$, we would have a contradiction since f is one-to-one; therefore $K(f) = \{e\}$.

Since an isomorphism is one-to-one and onto, it has an inverse which is a function from H onto G. The next theorem shows that the inverse is also an isomorphism.

Theorem 3.5: If f: $G \rightarrow H$ is an isomorphism, then f^{-1}: $H \rightarrow G$ is an isomorphism.

Proof: Since f is one-to-one and onto, it follows that f^{-1} is also one-to-one and onto. Let a_0 be the element of H such that $a = f^{-1}(a_0)$ for any $a \in G$; then $a * b = f^{-1}(a_0) * f^{-1}(b_0)$. But $a * b$ is the inverse image of the element $a_0 \circ b_0 \in H$ because $f(a * b) = f(a) \circ f(b) = a_0 * b_0$ since f is a homomorphism. Therefore $f^{-1}(a_0) * f^{-1}(b_0) = f^{-1}(a_0 \circ b_0)$, which shows that f^{-1} satisfies the definition (3.1) of a homomorphism.

Theorem 3.6: A homomorphism f: $G \rightarrow H$ is an isomorphism if it is onto and if its kernel contains only the identity element of G.

The proof of this theorem is a trivial consequence of Theorem 3.4 and the definition of an isomorphism.

Exercises

1. If f: $G \rightarrow H$ and $g : H \rightarrow M$ are homomorphisms, then show that the composition of the mappings f and g is a homomorphism from G to M.
2. Prove Theorems 3.3 and 3.6.
3. Show that the logarithm function is an isomorphism from the positive real numbers under multiplication to all the real numbers under addition. What is the inverse of this isomorphism?
4. Show that the function f: f:$(R,+) \rightarrow (R^+,\cdot)$ defined by $f(x) = x^2$ is not a homomorphism.

Rings and Fields

Groups and semi-groups are important building blocks in developing algebraic structures. In this section, we shall consider sets that form a group with respect to one binary operation and a semi-group with respect to another. We shall also consider a set that is a group with respect to two different binary operations.

Definition: A *ring* is a triple $(D,+,\cdot)$ consisting of a set D and two binary operations + and $\cdot$ such that

(a) D with the operation + is an Abelian group.

(b) The operation $\cdot$ is associative.

(c) D contains an identity element, denoted by 1, with respect to the operation $\cdot$, i.e.,

$$1 \cdot a = a \cdot 1 = a$$

for all $a \in D$.

(d) The operations + and $\cdot$ satisfy the distributive axioms

$$a .(b + c) = a . b + a . c$$

$$(b + c) . a = b . a + c . a \tag{3.6}$$

The operation + is called addition and the operation $\cdot$ is called *multiplication*. As was done with the notation for a group, the ring $(D,+,\cdot)$ will be written simply as D. Axiom (a) requires that D contains the identity element for the + operation. We shall follow the usual procedure and denote this element by 0. Thus, $a + 0 = 0 + a = a$. Axiom (a) also requires that each element $a \in D$ have an additive inverse, which we will denote by $-a$, $a + (-a) = 0$. The quantity $a + b$ is called the *sum* of a and b, and the difference between a and b is $a +(-b)$, which is usually written as $a - b$. Axiom (b) requires the set D and the multiplication operation to form a semi-group. If the multiplication operation is also commutative, the ring is called a *commutative ring*. Axiom (c) requires that D contain an identity element for multiplication; this element is called the unity element of the ring and is denoted by 1. The symbol 1 should not be confused with the real number one. The existence of the unity element is sometimes omitted in the ring axioms and the ring as we have defined it above is called the *ring with unity*. Axiom (d), the distributive axiom, is the only idea in the definition of a ring that has not appeared before. It provides a rule for the interchanging of the two binary operations.

The following familiar systems are all examples of rings.

1. The set I of integers with the ordinary addition and multiplication operations form a ring.
2. The set R_a of rational numbers with the usual addition and multiplication operations form aa ring.
3. The set R of real numbers with the usual addition and multiplication operations form a ring.
4. The set C of complex numbers with the usual addition and multiplication operations form a ring.
5. The set of all N by N matrices form a ring with respect to the operations of matrix addition and matrix multiplication. The unity element is the unity matrix and the zero element is the zero matrix.
6. The set of all polynomials in one (or several) variable with real (or complex) coefficients form a ring with respect to the usual operations of addition and multiplication.

Many properties of rings can be deduced from similar results that hold for groups. For example, the zero element 0, the unity element 1, the negative element $-a$ of an element a are unique. Properties that are associated with the interconnection between the two binary operations are contained in the following theorems:

Theorem 3.7: For any element $a \in D$, $a \cdot 0 = 0 \cdot a = 0$.

Proof: For any $b \in D$ we have by Axiom (a) that $b + 0 = b$. Thus for any $a \in D$ it follows that $(b + 0) \cdot a = b \cdot a$ and the Axiom (d) permits this equation to be recast in the form $b \cdot a + 0 \cdot a = b \cdot a$. From Property 8 for the additive group this equation implies that $0 \cdot a = 0$. It can be shown that $a \cdot 0 = 0$ by a similar argument.

Theorem 3.8: For all elements $a,b \in D$

$$(-a) \cdot b = a \cdot (-b) = -(a \cdot b)$$

This theorem shows that there is no ambiguity in writing $-a \cdot b$ for $-(a \cdot b)$.

Many of the notions developed for groups can be extended to rings; for example, subrings and ring homomorphisms correspond to the notions of subgroups and group homomorphisms. The interested reader can consult the Selected Reading for a discussion of these ideas.

The set of integers is an example of an algebraic structure called an *integral domain.*

Definition: A ring D is an *integral domain* if it satisfies the following additional axioms:

(e) The operation $\cdot$ is commutative.

(f) If a,b,c are any elements of D with $c \neq 0$, then

$$a \,.\, c = b \,.\, c \Rightarrow a = b \tag{3.7}$$

The cancellation law of multiplication introduced by Axiom (f) is logically equivalent to the assertion that a product of non-zero factors is non-zero. This is proved in the following theorem.

Theorem 3.9: A commutative ring D is an integral domain if and only if for all elements $a,b \in D$, $a \cdot b \neq 0$ unless $a = 0$ or $b = 0$.

Proof: Assume first that D is an integral domain so the cancellation law holds. Suppose $a \cdot b = 0$ and $b \neq 0$. Then we can write $a \cdot b = 0 \cdot b$ and the cancellation law implies $a = 0$. Conversely, suppose that a product in a commutative ring D cannot be zero unless one of its factors is zero. Consider the expression $a \cdot c = b \cdot c$, which can be rewritten as $a \cdot c - b \cdot c = 0$ or as $(a - b) \cdot c = 0$. If $c \neq 0$ then by assumption $a - b = 0$ or $a = b$. This proves that the cancellation law holds.

The sets of integers I, rational numbers R_a, real numbers R, and complex numbers C are examples of integral domains as well as being examples of rings. Sets of square matrices, while forming rings with respect to the binary operations of matrix addition and multiplication, do not, however, form integral domains. We can show this in several ways; first by showing that matrix multiplication is not commutative and, second, we can find two non-zero matrices whose product is zero, for example,

$$\begin{bmatrix} 5 & 0 \\ 0 & 0 \end{bmatrix}\begin{bmatrix} 0 & 0 \\ 0 & 1 \end{bmatrix} = \begin{bmatrix} 0 & 0 \\ 0 & 0 \end{bmatrix}$$

The rational numbers, the real numbers, and the complex numbers are examples of an algebraic structure called a *field.*

Definition: A field F is an integral domain, containing more than one element, and such that any element $a \in F/\{0\}$ has an inverse with respect to multiplication.

It is clear from this definition that the set F and the addition operation as well as the set $F/\{0\}$ and the multiplication operation form Abelian groups. Hence the unity element 1 , the zero element 0, the negative $(-a)$, as well as the reciprocal $(1/a)$, $a \neq 0$, are unique. The formulas of arithmetic can be developed as theorems following from the axioms of a field. It is not our purpose to do this here; however, it is important to convince oneself that it can be done.

The definitions of ring, integral domain, and field are each a step more restrictive than the other. It is trivial to notice that any set that is a field is automatically an integral domain and a ring. Similarly, any set that is an integral domain is automatically a ring.

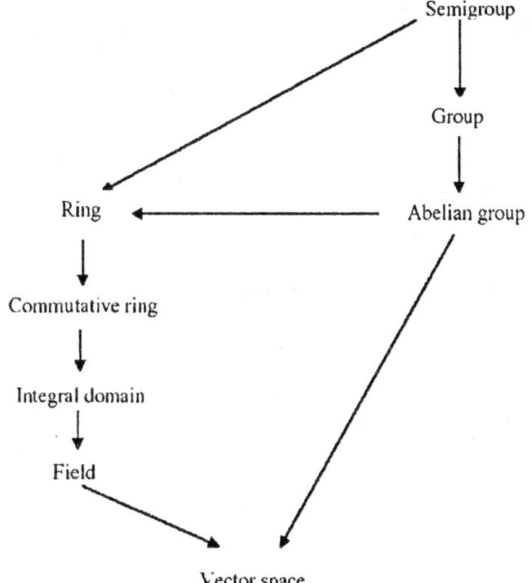

Figure: A Schematic of Algebraic Structures.

The dependence of the algebraic structures introduced in this chapter is illustrated schematically in Figure. The dependence of the vector space, which is to be introduced in the next chapter, upon these algebraic structures is also indicated.

Exercises

1. Verify that Examples 1 and 5 are rings.
2. Prove Theorem 3.8.
3. Prove that if a,b,c,d are elements of a ring D, then
   (a) $(-a) \cdot (-b) = a \cdot b$.
   (b) $(-1) \cdot a = a \cdot (-1) = -a$.
   (c) $(a + b) \cdot (c + d) = a \cdot c + a \cdot d + b \cdot c + b \cdot d$.
4. Among the Examples 1-6 of rings given in the text, which ones are actually integral domains, and which ones are fields?
5. Is the set of rational numbers a field?
6. Why does the set of integers not constitute a field?
7. If F is a field, why is $F/\{0\}$ an Abelian group with respect to multiplication?
8. Show that for all rational numbers x,y, the set of elements of form $x + y\sqrt{2}$ constitutes field.
9. For all rational numbers x,y,z, does the set of elements of the form $x + y\sqrt{2} + z\sqrt{3}$ form a field? If not, show how to enlarge it to a field.

Vector Spaces

Axioms for a Vector Space

Generally, one's first encounter with the concept of a vector is a geometrical one in the form of a directed line segment, that is to say, a straight line with an arrowhead. This type of vector, if it is properly defined, is a special example of the more general notion of a vector presented in this chapter. The concept of a vector put forward here is purely algebraic. The definition given for a vector is that it be a member of a set that satisfies certain algebraic rules.

A vector space is a triple (V, F, f) consisting of (a) an additive Abelian group V, (b) a field F, and (c) a function $f: F \times V \to V$ called *scalar multiplication* such that

$$\begin{aligned} f(\lambda, f(\mu, v)) &= f(\lambda\mu, v) \\ f(\lambda+\mu, u) &= f(\lambda, u) + f(\mu, u) \\ f(\lambda, u+v) &= f(\lambda, u) + f(\lambda, v) \\ f(1, v) &= v \end{aligned} \tag{4.1}$$

for all $\lambda, \mu \in F$ and all $u,v \in V$. A vector is an element of a vector space. The notation (V, F, f) for a vector space will be shortened to simply V. The first of (4.1) is usually called the *associative law* for *scalar multiplication*, while the second and third equations are distributive law, the second for scalar addition and the third for vector addition.

It is also customary to use a simplified notation for the scalar multiplication function *f*. We shall write $f(\lambda, v) = \lambda v$ and also regard λv and $v\lambda$ to be identical. In this simplified notation, we shall now list in detail the axioms of a vector space. In this definition the vector u + v in *V* is called the *sum* of u and v and the difference of u and v is written u – v and is defined by

$$u - v = u + (-v) \tag{4.2}$$

Definition: Let *V* be a set and *F* a field. *V* is a vector space if it satisfies the following rules:

(a) There exists a binary operation in *V* called addition and denoted by + such that

(1) $(u + v) + w = u + (v + w)$ for all $u, v, w \in V$.

(2) $u + v = v + u$ for all $u, v \in V$.

(3) There exists an element $0 \in V$ such that $u + 0 = u$ for all $u \in V$.

(4) For every $u \in V$ there exists an element $-u \in V$. such that $u + (-u) = 0$.

(b) There exists an operation called *scalar multiplication* in which every scalar $\lambda \in F$ can be combined with every element $u \in V$ to give an element $\lambda u \in V$ such that

(1) $\lambda(\mu u) = (\lambda\mu)u$

(2) $(\lambda + \mu)u = \lambda u + \mu u$

(3) $\lambda(u + v) = \lambda u + \lambda v$

(4) $1u = u$

for all $\lambda, \mu \in F$ and all $u, v \in V$.

If the field *F* employed in a vector space is actually the field of real numbers *R*, the space is called a *real vector space*. A complex vector space is similarly defined.

Except for a few minor examples, the material in this chapter and the next three employs complex vector spaces. Naturally the real vector space is a trivial special case. The reason for allowing the scalar field to be complex in these chapters is that the material on spectral decompositions has more usefulness in the complex case. Therefore, unless we provide some qualifying statement to the contrary, a vector space should be understood to be a complex vector space.

There are many and varied sets of objects that qualify as vector spaces. The following is a list of examples of vector spaces:

1. The vector space C^N is the set of all *N*-tuples of the form u = $(\lambda_1, \lambda_2, ..., \lambda_N)$, where *N* is a positive integer and $\lambda_1, \lambda_2, ..., \lambda_N \in C$. Since a *N*-tuple is an ordered set, if v = $(\mu_1, \mu_2, ..., \mu_N)$ is a second *N*-tuple, then and u and v are equal if and only if

   $$\mu_k = \lambda_k \text{ for all } k = 1, 2, ..., N$$

   The zero N-tuple is 0 = (0, 0,...,0) and the negative of the *N*-tuple u is -u $(-\lambda_1, -\lambda_2, ..., -\lambda_N)$. Addition and scalar multiplication of *N*-tuples are defined by the formulas

   $$\mathrm{u} + \mathrm{v} = (\mu_1 + \lambda_1, \mu_2, + \lambda_2, ..., \mu_N + \lambda_N)$$

   and

   $$\mu\,\mathrm{u} = (\mu\lambda_1, \mu\lambda_2, ..., \mu\lambda_N)$$

   respectively. The notation C^N is used for this vector space because it can be considered to be an *N*th Cartesian product of *C*.

2. The set *V* of all *N* × *M* complex matrices is a vector space with respect to the usual operation of matrix addition and multiplication by a scalar.

3. Let *H* be a vector space whose vectors are actually functions defined on a set *A* with values in C. Thus, if $h \in H$, $x \in A$ then h(x) $\in$ C and h: A $\to$ C. If *k* is another vector of *H* then equality of vectors is defined by

   $$\mathrm{h} = \mathrm{k} \Leftrightarrow \mathrm{h}(x) = \mathrm{k}(x) \text{ for all } x \in A$$

The zero vector is the zero function whose value is zero for all x. Addition and scalar multiplication are defined by

$$(h + k)(x) = h(x) + k(x)$$

and

$$(\lambda h)(x) = \lambda(h(x))$$

respectively.

4. Let P denote the set of all polynomials u of degree N of the form

$$u = \lambda_0 + \lambda_1 x + \lambda_2 x^2 + \cdots + \lambda_N x^N$$

where $\lambda_0, \lambda_1, \lambda_2, \ldots, \lambda_N \in C$. The set P forms a vector space over the complex numbers C if addition and scalar multiplication of polynomials are defined in the usual way.

5. The set of complex numbers C, with the usual definitions of addition and multiplication by a real number, forms a real vector space.

6. The zero element 0 of any vector space forms a vector space by itself.

The operation of scalar multiplication is not a binary operation as the other operations we have considered have been. In order to develop some familiarity with the algebra of this operation consider the following three theorems.

Theorem 4.1: $\lambda u = 0 \Leftrightarrow \lambda = 0$ or $u = 0$.

Proof: The proof of this theorem actually requires the proof of the following three assertions:

(a) $0u = 0$, (b) $\lambda 0 = 0$, (c) $\lambda u = 0 \Rightarrow \lambda = 0$ or $u = 0$

To prove (a), take $\mu = 0$ in Axiom (b2) for a vector space; then $\lambda u = \lambda u + 0u$. Therefore

$$\lambda u - \lambda u = \lambda u - \lambda u + 0u$$

and by Axioms (a4) and (a3)

$$0 = 0 + 0u = 0u$$

which proves (a).

To prove (b), set v = 0 in Axiom (b3) for a vector space; then $\lambda u = \lambda u + \lambda 0$. Therefore

$$\lambda u - \lambda u = \lambda u - \lambda u + \lambda 0$$

and by Axiom (a4)

$$0 = 0 + \lambda 0 = \lambda 0$$

To prove (c), we assume $\lambda u = 0$. If $\lambda = 0$, we know from (a) that the equation $\lambda u = 0$ is satisfied. If $\lambda \neq 0$, then we show that *u* must be zero as follows:

$$u = 1u = \lambda\left(\frac{1}{\lambda}\right)u = \frac{1}{\lambda}(\lambda u) = \frac{1}{\lambda}(0) = 0$$

Theorem 4.2: $(-\lambda)$ u = $-\lambda$u for all $\lambda \in C$, $u \in V$.

Proof: Let $\mu = 0$ and replace λ by $-\lambda$ in Axiom (b2) for a vector space and this result follows directly.

Theorem 4.3: $-\lambda$u = $\lambda(-u)$ for all $\lambda \in C$, $u \in V$.

Finally, we note that the concepts of length and angle have not been introduced. The reason for the delay in their introduction is to emphasise that certain results can be established without reference to these concepts.

Exercises

1. Show that at least one of the axioms for a vector space is redundant. In particular, one might show that Axiom (a2) can be deduced from the remaining axioms.
   [*Hint*: expand (1 + 1)(u + v) by the two different rules].
2. Verify that the sets listed in Examples 1- 6 are actually vector spaces. In particular, list the zero vectors and the negative of a typical vector in each example.
3. Show that the axioms for a vector space still make sense if the field *F* is replaced by a ring. The resulting structure is called a *module over a ring*.

4. Show that the Axiom (b4) for a vector space can be replaced by the axiom

$$(b4)' \ \lambda u = 0 \Leftrightarrow \lambda = 0 \text{ or } u = 0$$

5. Let V and U be vector spaces. Show that the set $V \times U$ is a vector space with the definitions

$$(u, x) + (v, y) = (u + v, x + y)$$

and

$$\lambda(u, x) = (\lambda u, \lambda x)$$

where $u, v \in V$; $x, y \in U$; and $\lambda \in F$

6. Prove Theorem 4.3.

7. Let V, be a vector space and consider the set $V \times V$. Define addition in $V \times V$ by

$$(u, v) + (x, y) = (u + x, v + y)$$

and multiplication by complex numbers by

$$(\lambda + i\mu)(u, v) = (\lambda u - \mu v, \mu u + \lambda v)$$

where $\lambda, \mu \in R$. Show that $V \times V$ is a vector space over the field of complex numbers.

Independence, Dimension and Basis

The concept of linear independence is introduced by first defining what is meant by linear dependence in a set of vectors and then defining a set of vectors that is not linearly dependent to be linearly independent. The general definition of linear dependence of a set of N vectors is an algebraic generalisation and abstraction of the concepts of collinearity from elementary geometry.

Definition: A finite set of $N(N \geq 1)$ vectors $\{v_1, v_2, \ldots, v_N\}$ in a vector space V is said to be *linearly dependent* if there exists a set of scalars $\{\lambda^1, \lambda^2, \ldots, \lambda^N\}$, not all zero, such that

$$\sum_{j=1}^{N} \lambda^j v_j = 0 \tag{4.3}$$

The essential content of this definition is that at least one of the vectors $\{v_1, v_2,...,v_N\}$ can be expressed as a linear combination of the other vectors. This means that if $\lambda^1 \neq 0$, then $v_1 = \sum_{i=2}^{N} \mu^j v_j$, where $\mu^j = -\lambda^j/\lambda^1$ for j = 2,3,...,*N*. As a numerical example, consider the two vectors

$$v_1 = (1, 2, 3), v_2 = (3, 6, 9)$$

from R^3. These vectors are linearly dependent since

$$v_2 = 3v_1$$

The proof of the following two theorems on linear dependence is quite simple.

Theorem **4.4:** If the set of vectors $\{v_1, v_2, \cdots, v_N\}$ is linearly dependent, then every other finite set of vectors containing $\{v_1, v_2, \cdots, v_N\}$ is linearly dependent.

Theorem **4.5:** Every set of vectors containing the zero vector is linearly dependent.

A set of $(N \geq 1)$ vectors that is not linearly dependent is said to be *linearly independent*. Equivalently, a set of $N(N \geq 1)$ vectors $\{v_1, v_2, ..., v_N\}$ is linearly independent if (4.3) implies $\lambda^1 = \lambda^2 = ... = \lambda^N = 0$. As a numerical example, consider the two vectors v_1 = (1,2) and v_2 = (2,1) from R^2. These two vectors are linearly independent because

$$\lambda v_1 + \lambda v_2 = \lambda(1,2) + \mu(2,1) = 0 \Leftrightarrow \lambda + 2\mu = 0, 2\lambda + \mu = 0$$

and

$$\lambda + 2\mu = 0,\ 2\lambda + \mu = 0 \Leftrightarrow \lambda = 0,\ \mu = 0$$

Theorem **4.6:** Every non-empty subset of a linearly independent set is linearly independent.

A linearly independent set in a vector space is said to be *maximal* if it is not a proper subset of any other linearly independent set. A vector space that contains a (finite) maximal, linearly independent set is then said to be *finite dimensional.* Of course, if V is not finite dimensional, then it is called *infinite dimensional.* In this text, we shall be concerned only with finite-dimensional vector spaces.

Theorem 4.7: Any two maximal, linearly independent sets of a finite-dimensional vector space must contain exactly the same number of vectors.

Proof: Let $\{v_1,...,v_N\}$ and $\{u_1,....,u_M\}$ be two maximal, linearly independent sets of V. Then we must show that $N = M$. Suppose that $N \neq M$, say $N < M$. But the fact that $\{v_1,...,v_N\}$ is maximal, the sets $\{v_1,...,v_N, u_1\},...,\{v_1,...v_N, u_M\}$ are all linearly dependent. Hence there exist relations

$$\begin{aligned} \lambda_{11}v_1 + \cdots + \lambda_{1N}v_N + \mu_1 u_1 &= 0 \\ &\vdots \\ \lambda_{M1}v_1 + \cdots + \lambda_{MN}v_N + \mu_M u_M &= 0 \end{aligned} \tag{4.4}$$

where the coefficients of each equation are not all equal to zero. In fact, the coefficients $\mu_1,...,\mu_M$ of the vectors $u_1,...,u_M$ are all non-zero, for if $\mu_i = 0$ for any i, then

$$\lambda_{i1}v_1 + \cdots + \lambda_{iN}v_N = 0$$

for some non-zero $\lambda_{i1},...,\lambda_{iN}$ contradicting the assumption that $\{v_1,...,v_N\}$ is a linearly independent set. Hence we can solve (4.4) for the vectors $\{u_1,....,u_M\}$ in terms of the vectors $\{v_1,...,v_N\}$, obtaining

$$\begin{aligned} u_1 &= \mu_{11}v_1 + \cdots + \mu_{1N}v_N \\ u_M &= \mu_{M1}v_1 + \cdots + \mu_{MN}v_N \end{aligned} \tag{4.5}$$

where $\mu_{ij} = -\lambda_{ij} / \mu$ for $i = 1,...,M; j = 1,...,N$.

Now we claim that the first N equations in the above system can be inverted in such a way that the vectors $\{v_1,..., v_N\}$ are given by linear combinations of the vectors $\{u_1,....,u_N\}$. Indeed, inversion is possible if the coefficient matrix $\left[\mu_{ij}\right]$ for $i, j = 1,...,N$ is non-singular. But this is clearly the case, for if that matrix were singular, there would be non-trivial solutions $\{\alpha_1,\ldots,\alpha_N\}$ to the linear system

$$\sum_{i=1}^{N} \alpha_i \mu_{ij} = 0, \qquad j = 1,\ldots,N \tag{4.6}$$

Then from (4.5) and (4.6) we would have

$$\sum_{i=1}^{N} \alpha_i u_i = \sum_{j=1}^{N} \left(\sum_{i=1}^{N} \alpha_i \mu_{ij} \right) v_j = 0$$

contradicting the assumption that set $\{u_1,\ldots,u_N\}$, being a subset of the linearly independent set $\{u_1,\ldots,u_M\}$, is linearly independent.

Now if the inversion of the first N equations of the system (4.5) gives

$$\begin{aligned} v_1 &= \xi_{11} u_1 + \cdots + \xi_{1N} u_N \\ &\vdots \\ v_N &= \xi_{N1} u_1 + \cdots + \xi_{NN} u_N \end{aligned} \tag{4.7}$$

where $\left[\xi_{ij}\right]$ is the inverse of $\left[\mu_{ij}\right]$ for $i, j = 1,\ldots,N$, we can substitute (4.7) into the remaining $M - N$ equations in the system (4.5), obtaining

$$\begin{aligned} u_{N+1} &= \sum_{j=1}^{N} \left(\sum_{i=1}^{N} \mu_{N+1\,j} \xi_{ji} \right) u_i \\ &\vdots \\ u_M &= \sum_{j=1}^{N} \left(\sum_{i=1}^{N} \mu_{Mj} \xi_{ji} \right) u_i \end{aligned}$$

But these equations contradict the assumption that the set $\{u_1,\ldots,u_M\}$ is linearly independent. Hence $M > N$ is impossible and the proof is complete.

An important corollary of the preceding theorem is the following.

Theorem 4.8: Let $\{u_1,\ldots, u_N\}$ be a maximal linearly independent set in V, and suppose that $\{v_1,\ldots,v_N\}$ is given by (4.7). Then $\{v_1,\ldots, v_N\}$ is also a maximal, linearly independent set if and only if the coefficient matrix $\left[\xi_{ij}\right]$ in (4.7) is non-singular. In particular, if

$v_i = u_i$ for $i = 1,\ldots,k-1, k+1,\ldots, N$ but $v_k \neq u_k$, then $\{u_1,\ldots,u_{k-1}, v_k, u_{k+1},\ldots,u_N\}$ is a maximal, linearly independent set if and only if the coefficient ξ_{kk} in the expansion of u_k in terms of $\{v_1,\ldots, v_N\}$ (4.7) is non-zero.

From the preceding theorems, we see that the number N of vectors in a maximal, linearly independent set in a finite dimensional vector space V is an intrinsic property of V. We shall call this number N the *dimension* of V, written dim V, namely N =dim V, and we shall call any maximal, linearly independent set of V a basis of that space. Theorem 4.8 characterises all bases of V as soon as one basis is specified. A list of examples of bases for vector spaces follows:

1. The set of N vectors

$$\underbrace{\begin{matrix}(1,0,0,\ldots,0)\\(0,1,0,\ldots,0)\\(0,0,1,\ldots,0)\\\vdots\\(0,0,0,\ldots,1)\end{matrix}}_{N}$$

   is linearly independent and constitutes a basis for C^N, called the *standard basis.*

2. If $M^{2\times 2}$ denotes the vector space of all 2×2 matrices with elements from the complex numbers C, then the four matrices

$$\begin{bmatrix}1&0\\0&0\end{bmatrix}, \begin{bmatrix}0&1\\0&0\end{bmatrix}, \begin{bmatrix}0&0\\1&0\end{bmatrix}, \begin{bmatrix}0&0\\0&1\end{bmatrix}$$

   form a basis for $M^{2\times 2}$ called the *standard basis.*

3. The elements 1 and $i = \sqrt{-1}$ form a basis for the vector space C of complex numbers over the field of real numbers.

The following two theorems concerning bases are fundamental.

Theorem 4.9: If $\{e_1, e_2,\ldots,e_N\}$ is a basis for V, then *every* vector in V, has the representation

$$v = \sum_{j-1}^{N} \xi^j e_j \tag{4.8}$$

where $\left\{\xi^1, \xi^2, ..., \xi^N\right\}$ are elements of C which depend upon the vector $v \in V$.

The proof of this theorem is contained in the proof of Theorem 4.7.

Theorem 4.10: The N scalars $\left\{\xi^1, \xi^2, ..., \xi^N\right\}$ in (4.8) are unique.

Proof: As is customary in the proof of a uniqueness theorem, we assume a lack of uniqueness. Thus we say that v has two representations,

$$v = \sum_{j=1}^{N} \xi^j e_j, \qquad v = \sum_{j=1}^{N} \mu^j e_j,$$

Subtracting the two representations, we have

$$\sum_{j=1}^{N} \left(\xi^j - \mu^j\right) e_j = 0$$

and the linear independence of the basis requires that

$$\xi^j = \mu^j, \qquad j = 1, 2, ..., N$$

The coefficients $\left\{\xi^1, \xi^2, ..., \xi^N\right\}$ in the representation (4.8) are the components of v with respect to the basis $\{e_1, e_2, ..., e_N\}$. The representation of vectors in terms of the elements of their basis is illustrated in the following examples.

1. The vector space C^3 has a standard basis consisting of e_1, e_2, and e_3;

   $$e_1 = (1, 0, 0), \qquad e_2 = (0, 1, 0), \quad e_3 = (0, 0, 1)$$

   A vector $v = (2 + i, 7, 8 + 3i)$ can be written as

   $$v = (2 + i)e_1 + 7e_2 + (8 + 3i)e_3$$

2. The vector space of complex numbers C over the space of real numbers R has the basis $\{1, i\}$. Any complex number z can then be represented by

$$z = \mu + \lambda i$$

where $\mu, \lambda \in R$.

3. The vector space $M^{2\times 2}$ of all 2×2 matrices with elements from the complex numbers C has the standard basis

$$e_{11} = \begin{bmatrix} 1 & 0 \\ 0 & 0 \end{bmatrix}, \; e_{12} = \begin{bmatrix} 0 & 1 \\ 0 & 0 \end{bmatrix} \; e_{21} = \begin{bmatrix} 0 & 0 \\ 1 & 0 \end{bmatrix}, \; e_{22} = \begin{bmatrix} 0 & 0 \\ 0 & 1 \end{bmatrix}$$

Any 2×2 matrix of the form

$$v = \begin{bmatrix} \mu & \lambda \\ v & \xi \end{bmatrix}$$

where $\mu, \lambda, v, \xi \in C$, can then be represented by

$$v = \mu e_{11} + \lambda e_{12} + v e_{21} + \xi e_{22}$$

The basis for a vector space is not unique and the general rule of change of basis is given by Theorem 4.8. An important special case of that rule is made explicit in the following exchange theorem.

Theorem 4.11: If $\{e_1, e_2, \ldots, e_N\}$ is a basis for V and if $B = \{b_1, b_2, \ldots, b_k\}$ is a linearly independent set of $K (N \geq K)$ in V, then it is possible to exchange a certain K of the original base vectors with $b_1, b_2, \ldots, b_k$ so that the new set is a basis for V.

Proof: We select b_1 from the set B and order the basis vectors such that the component ξ^1 of b_1 is not zero in the representation

$$b_1 = \sum_{j=1}^{N} \xi^j e_j$$

By Theorem 4.8, the vectors $\{b_1, e_1, e_2, \ldots, e_N\}$ form a basis for V. A second vector b_2 is selected from V and we again order the basis vectors so that this time the component λ^2 is not zero in the formula

$$b_2 = \lambda^1 b_1 + \sum_{j=2}^{N} \lambda^j e_j$$

Again, by Theorem 4.8 the vectors $\{b_1, b_2, e_3, \ldots, e_N\}$ form a basis for V. The proof is completed by simply repeating the above construction $K - 2$ times.

We now know that when a basis for V is given, every vector in V has a representation in the form (4.8). Inverting this condition somewhat, we now want to consider a set of vectors $B = \{b_1, b_2, \ldots, b_M\}$ of V with the property that every vector $v \in V$ can be written as

$$v = \sum_{j=1}^{M} \lambda^j b_j$$

Such a set is called a *generating set* of V (or is said to *span* V). In some sense a generating set is a counterpart of a linearly independent set. The following theorem is the counter part of Theorem 4.4.

Theorem 4.12: If $B = \{b_1, \ldots, b_M\}$ is a generating set of V, then every other finite set of vectors containing B is also a generating set of V.

In view of this theorem, we see that the counterpart of a maximal, linearly independent set is a minimal generating set, which is defined by the condition that a generating set $\{b_1, \ldots, b_M\}$ is minimal if it contains no proper subset that is still a generating set. The following theorem shows the relation between a maximal, linearly independent set and a minimal generating set.

Theorem 4.13: Let $B = \{b_1, \ldots, b_M\}$ be a finite subset of a finite dimensional vector space V. Then the following conditions are equivalent:

(i) B is a maximal linearly independent set.

(ii) B is a linearly independent generating set.

(iii) B is a minimal generating set.

Proof: We shall show that (i) $\Rightarrow$ (ii) $\Rightarrow$ (iii) $\Rightarrow$ (i).

(i) $\Rightarrow$ (ii). This implication is a direct consequence of the representation (4.8).

(ii) $\Rightarrow$ (iii). This implication is obvious. For if B is a linearly independent generating set but not a minimal generating set, then we can remove at least one vector, say b_M, and the remaining set is still a generating set. But this is impossible because b_M, can then be expressed as a linear combination of $\{b_1, \ldots, b_{M-1}\}$, contradicting the linear independence of B.

(iii) $\Rightarrow$ (i). If B is a minimal generating set, then B must be linearly independent because otherwise one of the vectors of B, say b_M, can be written as a linear combination of the vectors $\{b_1, \ldots, b_{M-1}\}$. It follows then that $\{b_1, \ldots, b_{M-1}\}$ is still a generating set, contradicting the assumption that $\{b_1, \ldots, b_M\}$ is minimal. Now a linearly independent generating set must be maximal, for otherwise there exists a vector $b \in V$ such that $\{b_1, \ldots, b_M, b\}$ is linearly independent. Then b cannot be expressed as a linear combination of $\{b_1, \ldots, b_M\}$ thus contradicting the assumption that B is a generating set.

In view of this theorem, a basis B can be defined by any one of the three equivalent conditions (i), (ii), or (iii).

Exercises

1. In elementary plane geometry it is shown that two straight lines determine a plane if the straight lines satisfy a certain condition. What is the condition? Express the condition in vector notation.
2. Prove Theorems 4.4-4.6, and 4.12.
3. Let $M^{3\times3}$ denote the vector space of all 3×3 matrices with elements from the complex numbers C. Determine a basis for $M^{3\times3}$
4. Let $M^{2\times2}$ denote the vector space of all 2×2 matrices with elements from the real numbers R. Is either of the following sets a basis for $M^{2\times2}$?

$$\left\{\begin{bmatrix}1 & 0\\0 & 0\end{bmatrix} \begin{bmatrix}0 & 6\\0 & 2\end{bmatrix}, \begin{bmatrix}0 & 1\\3 & 0\end{bmatrix}, \begin{bmatrix}0 & 1\\6 & 8\end{bmatrix}\right\}$$

$$\left\{\begin{bmatrix}1 & 0\\0 & 0\end{bmatrix}, \begin{bmatrix}0 & 1\\0 & 0\end{bmatrix}, \begin{bmatrix}0 & 0\\0 & 1\end{bmatrix}, \begin{bmatrix}0 & 0\\0 & 0\end{bmatrix}\right\}$$

5. Are the complex numbers $2 + 4i$ and $6 + 2i$ linearly independent with respect to the field of real numbers R? Are they linearly independent with respect to the field of complex numbers?

Intersection, Sum and Direct Sum of Subspaces

In this section operations such as "summing" and "intersecting" vector spaces are discussed. We introduce first the important concept of a subspace of a vector space in analogy with a subgroup of a group. A non-empty subset U of a vector space V is a subspace if:

(a) $u, w \in U \Rightarrow u + w \in U$ for all $u, w \in U$.

(b) $u, \in U \Rightarrow \lambda u \in U$ for all $\lambda \in C$.

Conditions (a) and (b) in this definition can be replaced by the equivalent condition:

(a′) $u, w \in U \Rightarrow \lambda u + \mu w \in U$ for all $\lambda \in C$.

Examples of subspaces of vector spaces are given in the following list:

1. The subset of the vector space C^N of all N-tuples of the form $(0, \lambda_2, \lambda_3, ..., \lambda_N)$ is a subspace of C^N.
2. Any vector space V is a subspace of itself.
3. The set consisting of the zero vector $\{0\}$ is a subspace of V.
4. The set of real numbers R can be viewed as a subspace of the real space of complex numbers C.

The vector spaces $\{0\}$ and V itself are considered to be *trivial* subspaces of the vector space V. If U is not a trivial subspace, it is said to be a *proper subspace* of V.

Several properties of subspaces that one would naturally expect to be true are developed in the following theorems.

Theorem **4.14:** If U is a subspace of V, then $0 \in U$.

Proof: The proof of this theorem follows easily from (b) in the definition of a subspace above by setting $\lambda = 0$.

Theorem **4.15:** If U is a subspace of V, then dim $U \leq$ dim V.

Proof: By Theorem 4.11 we know that any basis of U can be enlarged to a basis of V; it follows that dim $U \leq$ dim V.

Theorem **4.16:** If U is a subspace of V, then dim U = dim V if and only if $U = V$.

Proof: If $U = V$, then clearly dim U = dim V. Conversely, if dim$U$ = dim V, then a basis for U is also a basis for V. Thus, any vector $v \in V$ is also in U, and this implies $U = V$.

Operations that combine vector spaces to form other vector spaces are simple extensions of elementary operations defined on sets. If U and W are subspaces of a vector space V the sum of U and W, written $U + W$, is the set

$$U = W = \{u + w \mid u \in U, w \in W\}$$

Similarly, if U and W are subspaces of a vector space V, the *intersection* of U and W, denoted by $U \cap W$, is the set

$$U \cap W = \{u \mid u \in U \text{ and } u \in W\}$$

The union of two subspaces U and W of V, denoted by $U \cup W$, is the set

$$U \cup W = \{u \mid u \in U \text{ or } u \in W\}$$

Some properties of these two operations are stated in the following theorems.

Theorem **4.17:** If U and W are subspaces of V, then $U + W$ is a subspace of V.

Theorem **4.18:** If U and W are subspaces of V, then U and W are also subspaces of $U + W$.

Theorem **4.19:** If U and W are subspaces of V, then the intersection $U \cap W$ is a subspace of.

Proof: Let $u, w \in U \cap W$. Then $u, w \in W$ and $u, w \in U$. Since U and W are subspaces, $u + w \in W$ and $u + w \in U$ which means that $u + w \in U \cap W$ also. Similarly, if $u \in U \cap W$, then for all $\lambda \in R$. $\lambda u \in U \cap W$.

Theorem 4.20: If U and W are subspaces of V, then the union $U \cup$ W is not generally a subspace of V.

Proof: Let $u \in U, u \notin W$, and let $w \in W$ and $w \notin U$; then $u + w \notin U$ and $u + w \notin W$, which means that $u + w \notin U \cup W$.

Theorem 4.21: Let v be a vector in U + W, where U and W are subspaces of V. The decomposition of $v \in U + W$ into the form $v = u + w$, where $u \in U$ and $w \in W$, is unique if and only if $U \cap W = \{0\}$.

Proof: Suppose there are two ways of decomposing v; for example, let there be a u and u' in U and a w and w' in W such that

$$v = u + w \text{ and } v = u' + w'$$

The decomposition of v is unique if it can be shown that the vector b,

$$b = u - u' = w' - w$$

vanishes. The vector b is contained in $U \cap W$ since u and u' are known to be in U, and w and w' are known to be in W. Therefore $U \cap W = \{0\}$ implies uniqueness. Conversely, if we have uniqueness, then $U \cap W = \{0\}$, for otherwise any non-zero vector $y \in U \cap W$ has at least two decompositions $y = y + 0 = 0 + y$.

The sum of two subspaces U and W is called the *direct sum* of U and W and denoted by $U \oplus W$ if $U \cap W = \{0\}$. This definition is motivated by the result proved in Theorem 4.21. If $U \oplus W = V$, then U is called the *direct complement* of W in V. The operation of direct summing can be extended to a finite number of subspaces V_1, V_2,..., V_Q of V. The direct sum $V_1 \oplus V_2 \oplus \cdots \oplus V_Q$ is required to satisfy the conditions

$$V_R \cap \sum_{K=1}^{K=R-1} V_K + V_R \cap \sum_{K=R+1}^{Q} V_K = \{0\} \text{ for } R = 1,2,\ldots,Q$$

The concept of a direct sum of subspaces is an important tool for the study of certain concepts in geometry. The following theorem shows that the dimension of a direct sum of subspaces is equal to the sum of the dimensions of the subspaces.

Theorem 4.22: If U and W are subspaces of V, then

$$\dim U \oplus W = \dim U + \dim W$$

Proof: Let $\{u_1, u_2,\ldots,u_R\}$ be a basis for U and $\{w_1, w_2,\ldots, w_Q\}$ be a basis for W.. Then the set of vectors $\{u_1, u_2,\ldots, u_R, w_1, w_2,\ldots, w_Q\}$ is linearly independent since $U \cap W = \{0\}$. This set of vectors generates $U \oplus W$ since for any $v \in U \oplus W$ we can write

$$v = u + w = \sum_{j=1}^{R} \lambda^j u_j + \sum_{j=1}^{Q} \mu^j w_j$$

where $u \in U$ and $w \in W$. Therefore by Theorem 4.13, $\dim U \oplus W = R + Q$.

The result of this theorem can easily be generalised to

$$\dim\left(V_1 \oplus V_2 \oplus \cdots \oplus V_Q\right) = \sum_{j=1}^{Q} \dim V_i$$

The designation "direct sum" is sometimes used in a slightly different context. If V and U are vector spaces, not necessarily subspaces of a common vector space, the set $V \times U$ can be given the vector space structure by defining addition and scalar multiplication. The set $V \times U$ with this vector space structure is also called the direct sum and is written $V \oplus U$. This concept of direct sum is slightly more general since V and U need not initially be subspaces of a third vector space. However, after we have defined $U \oplus V$, then U and V can be viewed as subspaces of $U \oplus V$; further, in that sense, $U \oplus V$ is the direct sum of U and V in accordance with the original definition of the direct sum of subspaces.

Exercises

1. Under what conditions can a single vector constitute a subspace?
2. Prove Theorems 4.17 and 4.18.
3. If U and W are subspaces of V, show that any subspace which contains $U \cup W$ also contains $U + W$.
4. If V and U are vector spaces and if $V \oplus U$ is their direct sum in the sense of the second definition given in this section, reprove Theorem 4.22.
5. Let V - R^3, and suppose that U is the subspace spanned by $\{(0,1,1)\}$ and W is the subspace spanned by $\{(1,0,1), (1,1,1)\}$. Show that $V = U \oplus W$. Find another subspace $\overline{U}$ such that $V = \overline{U} \oplus W$.

Factor Space

In this section an equivalence relation is introduced and the vector space is partitioned into equivalence sets. The class of all equivalence sets is itself a vector space called the *factor space.*

If U is a subset of a vector space V, then two vectors w and v in V are said to be equivalent with respect to U, written $v \sim w$, if $w - v$ is a vector contained in U. It is easy to see that this relation is an equivalence relation and it induces a partition of V into equivalence sets of vectors. If $v \in V$, then the equivalence set of v, denoted by $\overline{v}$,[1] is the set of all vectors of the form $v + u$, where u is any vector of U,

$$\overline{v} = \{v + u \mid u \in U\}$$

To illustrate the equivalence relation and its decomposition of a vector space into equivalence sets, we will consider the real vector space R^2, which we can represent by the Euclidean plane. Let u be a fixed vector in R^2, and define the subspace U of R^2 by $\{\lambda u \mid \lambda \in R\}$.

This subspace consists of all vectors of the form $\lambda u, \lambda \in R$, which are all parallel to the same straight line. This subspace is illustrated in in the following figure. From the definition of equivalence, the vector v is seen to be equivalent to the vector w if the vector $v - w$ is parallel to the line representing U. Therefore all vectors that differ from the vector v by a vector that is parallel to the line representing U are equivalent. The set of all vectors equivalent to the vector v is therefore the set of all vectors that terminate on the dashed line parallel to the line representing U in the figure below.

The equivalence set of v, $\bar{v}$, is the set of all such vectors. The following theorem is a special case and shows that an equivalence relation decomposes V into disjoint sets, that is to say, each vector is contained in one and only one equivalence set.

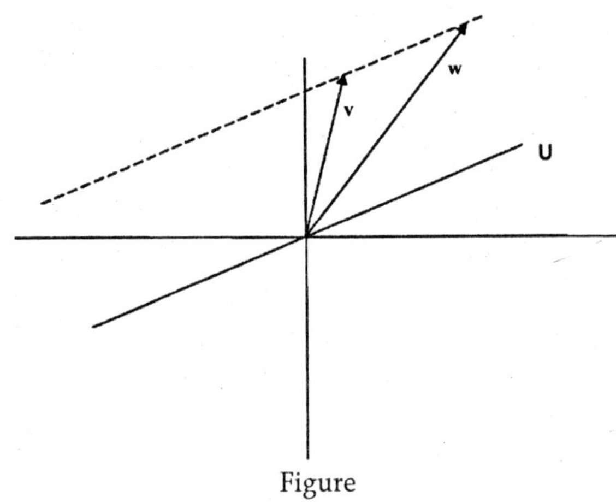

Figure

Theorem 4.23: If $\bar{v} \neq \bar{w}$, then $\bar{v} \cap \bar{w} = \varnothing$.

Proof: Assume that $\bar{v} \neq \bar{w}$ but that there exists a vector x in $\bar{v} \cap \bar{w}$. Then $x \in \bar{v}, x \sim v$, and $x \in \bar{w}, x \sim w$. By the transitive property of the equivalence relation we have $v \sim w$, which implies $\overline{v} = \overline{w}$ and which is a contradiction. Therefore $\bar{v} \cap \bar{w}$ contains no vector unless $\bar{v} = \bar{w}$, in which case $\bar{v} \cap \bar{w} = \bar{v}$.

We shall now develop the structure of the factor space. The factor class of U, denoted by V / U is the class of all equivalence sets in V formed by using a subspace U of V. The factor class is sometimes called a *quotient class*. Addition and scalar multiplication of equivalence sets are denoted by

$$\bar{v} + \bar{w} = \overline{v + w}$$

and

$$\lambda \bar{v} = \overline{\lambda v}$$

respectively. It is easy to verify that the addition and multiplication operations defined above depend only on the equivalence sets and not on any particular vectors used in representing the sets. The following theorem is easy to prove.

Theorem **4.24:** The factor set V/U forms a vector space, called a *factor space*, with respect to the operations of addition and scalar multiplication of equivalence classes defined above.

The factor space is also called a *quotient space*. The subspace U in V/U plays the role of the zero vector of the factor space. In the trivial case when $U = V$, there is only one equivalence set and it plays the role of the zero vector. On the other extreme when $U = \{0\}$, then each equivalence set is a single vector and $V/\{0\} = V$.

Exercises

1. Show that the relation between two vectors v and $w \in V$ that makes them equivalent is, in fact, an equivalence relation.
2. Give a geometrical interpretation of the process of addition and scalar multiplication of equivalence sets in R^2 in the previous figure.
3. Show that V is equal to the union of all the equivalence sets in V. Thus V/U is a class of non-empty, disjoint sets whose union is the entire space V.
4. Prove Theorem 4.24.
5. Show that dim (V/U) = dim V -dim U.

Inner Product Spaces

There is no concept of length or magnitude in the definition of a vector space we have been employing. The reason for the delay in the introduction of this concept is that it is not needed in many of the results of interest. To emphasise this lack of dependence on the concept of magnitude, we have delayed its introduction to this point. Our intended applications of the theory, however, do employ it extensively.

We define the concept of length through the concept of an inner product. An inner product on a complex vector space V is a function f: $V \times V \to C$ with the following properties:

1. $f(u,v) = \overline{f(v,u)}$;
2. $\lambda f(u,v) = f(\lambda u,v)$;
3. $f(u+w,v) = f(u,v) + f(w,v)$;
4. $f(u,u) \geq 0$ and $f(u,u) = 0 \Leftrightarrow u = 0$;

for all $u,v,w \in V$ and $\lambda \in C$. In Property 1 the bar denotes the complex conjugate. Properties 2 and 3 require that f be linear in its first argument; i.e., $f(\lambda u + \mu v, w) = \lambda f(u,w) + \mu f(v,w)$ for all $u,v,w \in V$ and all $\lambda, \mu \in C$. Property 1 and the linearity implied by Properties 2 and 3 insure that f is conjugate linear in its second argument; i.e., $f(u, \lambda v + \mu w) = \bar{\lambda} f(u,v) + \bar{\mu} f(u,w)$ for all $u,v,w \in V$ and all $\lambda, \mu \in C$. Since Property 1 ensures that f (u, v) is real, Property 4 is meaningful, and it requires that f be positive definite. There are many notations for the inner product. We shall employ the notation of the "dot product" and write

$$f(\mathrm{u}, v) = \mathrm{u} \cdot \mathrm{v}$$

An inner product space is simply a vector space with an inner product. To emphasise the importance of this idea and to focus simultaneously all its details, we restate the definition as follows.

Definition: A *complex inner product space,* or simply an *inner product space,* is a set V and a field C such that:

(a) There exists a binary operation in V called addition and denoted by + such that:

(1) $(u + v) + w = u + (v + w)$ for all $u,v,w \in V$.

(2) $u + v = v + u$ for all $u,v, \in V$.

(3) There exists an element $0 \in V$ such that $u + 0 = u$ for all $u \in V$.

(4) For every $u \in V$ there exists an element $-u \in V$ such that $u + (-u) = 0$.

(b) There exists an operation called *scalar multiplication* in which every scalar $\lambda \in C$ can be combined with every element $u \in V$ to give an element $\lambda u \in V$ such that:

(1) $\lambda(\mu u) = (\lambda\mu)u$;

(2) $(\lambda + \mu)u = \lambda u + \mu u$;

(3) $\lambda(u + v) = \lambda u + \lambda v$;

(4) $1u = u$;

for all $\lambda, u, \in C$ and all $u, v \in V$;

(c) There exists an operation called inner product by which any ordered pair of vectors u and v in V determines an element of C denoted by $u \cdot v$ such that

(1) $u \cdot v = \overline{v \cdot u}$;

(2) $\lambda u \cdot v = (\lambda u) \cdot v$;

(3) $(u + w) \cdot v = u \cdot v + w \cdot v$;

(4) $u \cdot u \geq 0$ and $u \cdot u = 0 \Leftrightarrow u = 0$;

for all $u, v, w \in V$ and $\lambda \in C$.

A real inner product space is defined similarly. The vector space C^N becomes an inner product space if, for any two vectors $u, v \in C^N$. where $u = (\lambda_1, \lambda_2, \ldots, \lambda_N)$ and $v = (\mu_1, \mu_2, \ldots, \lambda_N)$, we define the inner product of u and v by

$$u \cdot v = \sum_{j=1}^{N} \lambda_j \overline{\mu_j}$$

The *length* of a vector is an operation, denoted by $\| \ \|$, that assigns to each non-zero vector $v \in V$ a positive real number by the following rule:

$$\|v\| = \sqrt{v \cdot v} \tag{4.9}$$

Of course the length of the zero vector is zero. The definition represented by (4.9) is for an inner product space of N dimensions in general and therefore generalises the concept of "length" or "magnitude" from elementary Euclidean plane geometry to N-dimensional spaces. Before continuing this process of algebraically generalising geometric notions, it is necessary to pause and prove two inequalities that will aid in the generalisation process.

***Theorem* 4.25:** The *Schwarz inequality*,

$$|u \cdot v| \leq \| u \| \| v \| \tag{4.10}$$

is valid for any two vectors u, v in an inner product space.

Proof: The Schwarz inequality is easily seen to be trivially true when either u or v is the *0* vector, so we shall assume that neither u nor v is zero. Construct the vector $(u \cdot u)\, v - (v \cdot u)\, u$ and employ Property (c4), which requires that every vector have a non-negative length, hence

$$\left(\| u \|^2 \| v \|^2 - (u \cdot v)\left(\overline{v \cdot v}\right)\right) \| u \|^2 \geq 0$$

Since u must not be zero, it follows that

$$\| u \|^2 \| v \|^2 \geq (u \cdot v)\left(\overline{u \cdot v}\right) = | u \cdot v |^2$$

and the positive square root of this equation is Schwarz's inequality.

***Theorem* 4.26:** The triangle inequality

$$\| u + v \| \leq \| u \| + \| v \| \tag{4.11}$$

is valid for any two vectors u, v in an inner product space.

Proof: The squared length of $u + v$ can be written in the form

$$\| u + v \|^2 = (u + v) \cdot (u + v) = \| u \|^2 + \| v \|^2 + u \cdot v + v \cdot u$$

$$= \| u \|^2 + \| v \|^2 + 2\,\mathrm{Re}(u \cdot v)$$

where Re signifies the real part. By use of the Schwarz inequality this can be rewritten as

$$\| u + v \|^2 \leq \| u \|^2 + \| v \|^2 + 2 \| u \| \| v \| = (\| u \| + \| v \|)^2$$

Taking the positive square root of this equation, we obtain the triangular inequality.

For a real inner product space the concept of angle is defined as follows. The angle between two vectors u and v denoted by θ, is defined by

$$\cos\theta = \frac{u \cdot v}{\| u \| \| v \|} \tag{4.12}$$

This definition of a real-valued angle is meaningful because the Schwarz inequality in this case shows that the quantity on the right-hand side of (4.12) must have a value lying between 1 and -1, i.e.,

$$-1 \le \frac{u \cdot v}{\| u \| \| v \|} \le +1$$

Returning to complex inner product spaces in general, we say that two vectors u and v are orthogonal if $u \cdot v$ or $v \cdot u$ is zero. Clearly, this definition is consistent with the real case, since orthogonality then means $|\theta| = \pi / 2$.

The inner product space is a very substantial algebraic structure and parts of the structure can be given slightly different interpretations. In particular it can be shown that the length $\| v \|$ of a vector $v \in V$ is a particular example of a mathematical concept known as a norm, and thus an inner product space is a *normal space*. A norm on V is a real-valued function defined on V whose value is denoted by $\| v \|$ and which satisfies the following axioms:

(1) $\| v \| \ge 0$ and $\| v \| = 0 \Leftrightarrow v = 0$;

(2) $\| \lambda v \| = | \lambda | \| v \|$;

(3) $\| u \| + \| v \| \ge \| u + v \|$;

for all $u, v \in V$ and all $\lambda \in C$. In defining the norm of v, we have employed the same notation as that for the length of v because we will show that the length defined by an inner product is a norm, but the converse need not be true.

Theorem 4.27: The operation of determining the length $\| v \|$ of $v \in V$ is a norm on V.

Proof: The proof follows easily from the definition of length. Properties (c1), (c2), and (c4) of an inner product imply Axioms 1 and 2 of a norm, and the triangle inequality is proved by Theorem 4.26.

We will now show that the inner product space is also a metric space. A metric space is a non-empty set M equipped with a positive real-valued function $M \times M \to R$, called the *distance* function that satisfies the following axioms:

(1) $d(u,v) \geq 0$ *and* $d(u,v) = 0 \Leftrightarrow u = v$;

(2) $d(u,v) = d(v,u)$;

(3) $d(u,w) \leq d(u,v) + d(v,w)$;

for all $u,v,w \in M$. In other words, a metric space is a set M of objects and a positive-definite, symmetric distance function d satisfying the triangle inequality. The boldface notation for the elements of M is simply to save the introduction of yet another notation.

Theorem 4.28: An inner product space V is a metric space with the distance function given by

$$d(u,v) = \| u - v \| \tag{4.13}$$

Proof: Let $w = u - v$; then from the requirement that $\| w \| \geq 0$ and $\| w \| = 0 \Leftrightarrow w = 0$ it follows that

$$\| u - v \| \geq 0 \text{ and } \| u - v \| = 0 \Leftrightarrow u = v$$

Similarly, from the requirement that $\| w \| = \| -w \|$, it follows that

$$\| u - v \| = \| v - u \|$$

Finally, let u be replaced by $u - v$ and v by $v - w$ in the triangle inequality (4.11); then

$$\| u - w \| \leq \| u - v \| + \| v - w \|$$

which is the third and last requirement for a distance function in a metric space.

The inner product as well as the expressions (4.9) for the length of a vector can be expressed in terms of any basis and the components of the vectors relative to that basis. Let $\{e_1, e_2,..., e_N\}$ be a basis of V and denote the inner product of any two base vectors by e_{jk},

$$e_{jk} \equiv e_j \cdot e_k \equiv \overline{e_{kj}} \tag{4.14}$$

Thus, if the vectors u and v have the representations

$$u = \sum_{j=1}^{N} \lambda^j e_j, \quad v = \sum_{k=1}^{N} \mu^k e_k \tag{4.15}$$

relative to the basis $\{e_1, e_2,...,e_N\}$ then the inner product of u and v is given by

$$u \cdot v = \sum_{j=1}^{N} \lambda^j e_j \cdot \sum_{k=1}^{N} \mu^k e_k = \sum_{j=1}^{N}\sum_{k=1}^{N} \lambda^j \overline{\mu}^k e_{jk} \tag{4.16}$$

Equation (4.16) is the component expression for the inner product.

From the definition (4.9) for the length of a vector v and from (4.14) and (4.15), we can, write

$$\| v \| = \left(\sum_{j=1}^{N}\sum_{k=1}^{N} e_{jk} \mu^j \overline{\mu}^k \right)^{1/2} \tag{4.17}$$

This equation gives the component expression for the length of a vector. For a real inner product space, it easily follows that

$$\cos\theta = \frac{\sum_{j=1}^{N}\sum_{k=1}^{N} e_{jk} \mu^j \lambda^k}{\left(\sum_{p=1}^{N}\sum_{r=1}^{N} e_{pr} \mu^p \mu^r \right)^{1/2} \left(\sum_{l=1}^{N}\sum_{s=1}^{N} e_{sl} \lambda^s \lambda^l \right)^{1/2}} \tag{4.18}$$

In formulas (4.16)-(4.18) notice that we have changed the particular indices that indicate summation so that no more than two indices occur in any summand. The reason for changing these dummy indices of summation can be made apparent by failing to change them. For example, in order to write (4.17) in its present form, we can write the component expressions for v in two equivalent ways,

$$v = \sum_{j=1}^{N} \mu^j e_j, \quad v = \sum_{k=1}^{N} \mu^k e_k$$

and then take the dot product. If we had used the same indices to represent summation in both cases, then that index would occur four times in (4.17) and all the cross terms in the inner product would be left out.

Exercises

1. Derive the formula

   $$2u \cdot v = \| u + v \|^2 + i \| u + iv \|^2 - (1+i) \| u \|^2 - (1+i) \| v \|^2$$

   which expresses the inner product of two vectors in terms of the norm. This formula is known as the *polar identity*.

2. Show that the norm $\| \; \|$ induced by an inner product according to the definition (4.9) must satisfy the following *parallelogram law*;

   $$\| u + v \|^2 + \| u - v \|^2 = 2 \| v \|^2 + 2 \| u \|^2$$

   for all vectors u and v. Prove by counterexample that a norm in general need not be an induced norm of any inner product.

3. Use the definition of angle given in this section and the properties of the inner product to derive the law of cosines.

4. If V and U are inner produce spaces, show that the equation

   $$f((v,w),(u,b)) = v \cdot u + w \cdot b$$

   where $v,u \in V$ and $w,b \in U$, defines an inner product on $V \oplus U$.

5. Show that the only vector in V that is orthogonal to every other vector in V is the zero vector.
6. Show that the Schwarz and triangle inequalities become equalities if and only if the vectors concerned are linearly dependent.
7. Show that $\| u_1 - u_2 \| \geq \| u_1 \| - \| u_2 \|$ for all u_1, u_2 in an inner product space V.
8. Prove by direct calculation that $\sum_{j=1}^{N} \sum_{k=1}^{N} e_{jk} \mu^j \bar{\mu}^k$ in (4.17) is real.
9. If U is a subspace of an inner product space V prove that U is also an inner product space.
10. Prove that the $N \times N$ matrix $[e_{jk}]$ defined by (4.14) is non-singular.

Orthonormal Bases and Orthogonal Complements

Experience with analytic geometry tells us that it is generally much easier to use bases consisting of vectors which are orthogonal and of unit length rather than arbitrarily selected vectors. Vectors with a magnitude of 1 are called *unit vectors* or *normalised vectors.* A set of vectors in an inner product space V is said to be an orthogonal set if all the vectors in the set are mutually orthogonal, and it is said to be an orthonormal set if the set is orthogonal and if all the vectors are unit vectors. In equation form, an orthonormal set $\{i_1, \ldots, i_M\}$ satisfies the conditions

$$i_j \cdot i_k = \delta_{jk} \equiv \begin{cases} 1 & \text{if } j = k \\ 0 & \text{if } j \neq k \end{cases} \tag{4.19}$$

The symbol δ_{jk} introduced in the equation above is called the *Kronecker delta.*

Theorem 4.29: An orthonormal set is Linearly Independent.

Proof: Assume that the orthonormal set $\{i_1, i_2, \ldots, i_M\}$ is linearly dependent, that is to say there exists a set of scalars $\left\{\lambda^1, \lambda^2, \ldots, \lambda^M\right\}$, not all zero, such that

$$\sum_{j=1}^{M} \lambda^j i_j = 0$$

The inner product of this sum with the unit vector i_k gives the expression

$$\sum_{j=1}^{M} \lambda^j \delta_{jk} = \lambda^1 \delta_{1k} + \cdots + \lambda^M \delta_{MK} = \lambda^k = 0$$

for $k = 1,2,\ldots, M$. A contradiction has therefore been achieved and the theorem is proved.

As a corollary to this theorem, it is easy to see that an orthonormal set in an inner product space V can have no more than $N = \dim V$ elements. An orthonormal set is said to be complete in an inner product space V if it is not a proper subset of another orthonormal set in the same space. Therefore every orthonormal set with $N = \dim V$ elements is complete and hence maximal in the sense of a linearly independent set. It is possible to prove the converse: Every complete orthonormal set in an inner product space V has N = dim V elements. The same result is put in a slightly different fashion in the following theorem.

Theorem 4.30: A complete orthonormal set is a basis for V; such a basis is called an *orthonormal basis.*

Any set of linearly independent vectors can be used to construct an orthonormal set, and likewise, any basis can be used to construct an orthonormal basis. The process by which this is done is called the Gram-Schmidt orthogonalisation process and it is developed in the proof of the following theorem.

Theorem 4.31: Given a basis $\{e_1, e_2,\ldots,e_N\}$ of an inner product space V, then there exists an orthonormal basis $\{i_1,\ldots,i_N\}$ such that $\{e_1\ldots,e_k\}$ and $\{i_1,\ldots,i_k\}$ generate the same subspace U_k of V, for each $k = 1,\ldots, N$.

Proof: The construction proceeds in two steps; first a set of orthogonal vectors is constructed, then this set is normalised. Let

$\{d_1, d_2,...,d_N\}$ denote a set of orthogonal, but not unit, vectors. This set is constructed from $\{e_1, e_2,..., e_N\}$ as follows: Let $d_1 = e_1$ and put

$$d_2 = e_2 + \xi d_1$$

The scalar ξ will be selected so that d_2 is orthogonal to d_1; orthogonality of d_1 and d_2 requires that their inner product be zero; hence

$$d_2 \cdot d_1 = 0 = e_2 \cdot d_1 + \xi d_1 \cdot d_1$$

implies

$$\xi = -\frac{e_2 \cdot d_1}{d_1 \cdot d_1}$$

where $d_1 \cdot d_1 \neq 0$ since $d_1 \neq 0$. The vector d_2 is not zero, because e_1 and e_2 are linearly independent. The vector d_3 is defined by

$$d_3 = e_3 + \xi^2 d_2 + \xi^1 d_1$$

The scalars ξ^2 and ξ^1 are determined by the requirement that d_3 be orthogonal to both d_1 and d_2; thus

$$d_3 \cdot d_1 = e_3 \cdot d_1 + \xi^1 d_1 \cdot d_1 = 0$$

$$d_3 \cdot d_2 = e_3 \cdot d_2 + \xi^2 d_2 \cdot d_2 = 0$$

and, as a result,

$$\xi^1 = -\frac{e_1 \cdot d_1}{d_1 \cdot d_1}, \; \xi^2 = -\frac{e_3 \cdot d_2}{d_2 \cdot d_2}$$

The linear independence of e_1, e_2 and e_3 requires that d_3 be non-zero. It is easy to see that this scheme can be repeated until a set of N orthogonal vectors $\{d_1, d_2,...,d_N\}$ has been obtained. The orthonormal set is then obtained by defining

$$i_k = d_k \,/\, \| d_k \|, \; k = 1,2,...,N$$

It is easy to see that $\{e_1..., e_k\}$, $\{d_1,...d_k\}$, and, hence, $\{i_1,...,i_k\}$ generate the same subspace for each k.

The concept of mutually orthogonal vectors can be generalised to mutually orthogonal subspaces. In particular, if U is a subspace of an inner product space, then the orthogonal complement of U is a subset of V, denoted by $U^{\perp}$, such that

$$U^{\perp} = \{v \mid v \cdot u = 0 \text{ for all } u \in U\} \tag{4.20}$$

The properties of orthogonal complements are developed in the following theorem.

***Theorem* 4.32:** If U is a subspace of V, then (a) $U^{\perp}$ is a subspace of V and (b) $V = U \oplus U^{\perp}$.

Proof: (a) If u_1 and u_2 are in $U^{\perp}$, then

$$u_1 \cdot v = u_2 \cdot v = 0$$

for all $v \in U$. Therefore for any $\lambda_1, \lambda_2 \in C$,

$$(\lambda_1 u_1 + \lambda_2 u_2) \cdot v = \lambda_1 u_1 \cdot v + \lambda_2 u_2 \cdot v = 0$$

Thus $\lambda_1 u_1 + \lambda_2 u_2 \in U^{\perp}$.

(b) Consider the vector space $U + U^{\perp}$. Let $v \in U \cap U^{\perp}$. Then by (4.20), $v \cdot v = 0$, which implies $v = 0$. Thus $U + U^{\perp} = U \oplus U^{\perp}$. To establish that $V = U \oplus U^{\perp}$, let $\{i_1, \ldots, i_R\}$ be an orthonormal basis for U. Then consider the decomposition

$$v = \left\{ v - \sum_{q=1}^{R} (v \cdot i_q) i_q \right\} + \sum_{q=1}^{R} (v \cdot i_q) i_q$$

The term in the brackets is orthogonal to each i_q and it thus belongs to $U^{\perp}$. Also, the second term is in U Therefore $V = U + U^{\perp} = U \oplus U^{\perp}$, and, the proof is complete.

As a result of the last theorem, any vector $v \in V$ can be written uniquely in the form

$$v = u + w \tag{4.21}$$

where $u \in U$ and $w \in U^{\perp}$. If v is a vector with the decomposition indicated in (4.21), then

$$\|v\|^2 = \|u + w\|^2 = \|u\|^2 + \|w\|^2 + u \cdot w + w \cdot u = \|u\|^2 + \|w\|^2$$

It follows from this equation that

$$\|v\|^2 \geq \|u\|^2, \qquad \|v\|^2 \geq \|w\|^2$$

The equation is the well-known *Pythagorean theorem*, and the inequalities are special cases of an inequality known as *Bessel's inequality*.

Exercises

1. Show that with respect to an orthonormal basis the formulas (4.16)-(4.18) can be written as

$$u \cdot v = \sum_{j=1}^{N} \lambda^j \overline{\mu^j} \tag{4.22}$$

$$\|v\|^2 = \sum_{j=1}^{N} \mu^j \overline{\mu^j} \tag{4.23}$$

and, for a real vector space,

$$\cos\theta = \frac{\sum_{j=1}^{N} \lambda^j \mu^j}{\left(\sum_{p=1}^{N} \mu^p \mu^p\right)^{1/2} \left(\sum_{l=1}^{N} \mu^l \mu^l\right)^{1/2}} \tag{4.24}$$

2. Prove Theorem 4.30.

3. If U is a subspace of the inner product space V, show that

$$\left(U^{\perp}\right)^{\perp} = U$$

4. If V is an inner product space, show that

$$V^{\perp} = \{0\}$$

5. Given the basis

$$e_1 = (1, 1, 1),\ e_2 = (0, 1, 1),\ e_3 = (0, 0, 1)$$

for R^3, construct an orthonormal basis according to the Gram-Schmidt orthogonalisation process.

6. If U is a subspace of an inner product space V, show that

$$\dim U^{\perp} = \dim V - \dim U$$

7. Given two subspaces V_1 and V_2 of V, show that

$$(V_1 + V_2)^{\perp} = V_1^{\perp} \cap V_2^{\perp}$$

and

$$(V_1 \cap V_2)^{\perp} = V_1^{\perp} + V_2^{\perp}$$

8. In the proof of Theorem 4.32 we used the fact that if a vector is orthogonal to each vector i_q of a basis $\{i_1,\ldots,i_R\}$ of U, then that vector belongs to $U^{\perp}$. Give a proof of this fact.

Reciprocal Basis and Change of Basis

In this section we define a special basis, called the *reciprocal basis,* associated with each basis of the inner product space V. Components of vectors relative to both the basis and the reciprocal basis are defined and formulas for the change of basis are developed.

The set of N vectors $\{e^1, e^2,\ldots,e^N\}$ is said to be the reciprocal basis relative to the basis $\{e_1, e_2,\ldots, e_N\}$ of an inner product space V if

$$e^k \cdot e_s = \delta_s^k, \quad k,s = 1,2,\ldots,N \tag{4.25}$$

where the symbol δ_s^k is the Kronecker delta defined by

$$\delta_s^k = \begin{cases} 1 & k = s \\ 0, & k \neq s \end{cases} \tag{4.26}$$

Thus each vector of the reciprocal basis is orthogonal to $N - 1$ vectors of the basis and when its inner product is taken with the Nth vector of the basis, the inner product has the value one. The following two theorems show that the reciprocal basis just defined exists uniquely and is actually a basis for V.

Theorem 4.33: The reciprocal basis relative to a given basis exists and is unique.

Proof: *Existence.* We prove only existence of the vector e^1; existence of the remaining vectors can be proved similarly. Let U be the subspace generated by the vectors $e_2,..., e_N$ and suppose that $U^\perp$ is the orthogonal complement of U. Then dim $U = N - 1$, and from Theorem 4.22, dim $U^\perp = 1$. Hence we can choose a non-zero vector $w \in U^\perp$. Since $e_1 \notin U$, e_1 and w are not orthogonal,

$$e_1 \cdot w \neq 0$$

we can simply define

$$e^1 \equiv \frac{1}{e_1 \cdot w} w$$

Then e^1 obeys (4.25) for $k = 1$.

Uniqueness: Assume that there are two reciprocal bases, $\{e^1, e^2,...,e^N\}$ and $\{d^1, d^2,..., d^N\}$ relative to the basis $\{e_1, e_2,..., e_N\}$ of V. Then from (4.25)

$$e^k \cdot e_s = \delta_s^k \text{ and } d^k \cdot e_s = \delta_s^k \text{ for } k, s = 1, 2, ..., N$$

Subtracting these two equations, we obtain

$$\left(e^k - d^k\right) \cdot e_s = 0, \quad k, s = 1, 2, ..., N \tag{4.27}$$

Thus the vector $e^k - d^k$ must be orthogonal to $\{e_1, e_2,...,e_N\}$ Since the basis generates V, (4.27) is equivalent to

$$(e^k - d^k).\ \mathrm{v} = 0 \text{ for all } v \in W \tag{4.28}$$

In particular, we can choose v in (4.28) to be equal to $e^k - d^k$, and it follows then from the definition of an inner product space that

$$e^k = d^k, \quad k = 1, 2,..., N \tag{4.29}$$

Therefore the reciprocal basis relative to $\{e_1, e_2,..., e_N\}$ is unique.

The logic in passing from (4.28) to (4.29) has appeared before.

Theorem 4.34: The reciprocal basis $\{e^1, e^2,...,e^N\}$ with respect to the basis $\{e_1, e_2,...,e_N\}$ of the inner product space V is itself a basis for V.

Proof: Consider the linear relation

$$\sum_{q=1}^{N} \lambda_q e^q = 0$$

If we compute the inner product of this equation with e_k, k = 1,2,..., N, then

$$\sum_{q=1}^{N} \lambda_q e^q \cdot e_k = \sum_{q=1}^{N} \lambda_q \delta_k^q = \lambda_k = 0$$

Thus the reciprocal basis is linearly independent. But since the reciprocal basis contains the same number of vectors as that of a basis, it is itself a basis for V.

Since e^k, k = 1,2,..., N, is in V, we can always write

$$e^k = \sum_{q=1}^{N} e^{kq} e_q \tag{4.30}$$

where, by (4.25) and (4.14),

$$e^k \cdot e_s = \delta_s^k = \sum_{q=1}^{N} e^{kq} e_{qs} \tag{4.31}$$

and

$$e^k \cdot e^j = \sum_{q=1}^{N} e^{kq} \delta_q^j = e^{kj} = \overline{e^{jk}} \tag{4.32}$$

From a matrix viewpoint, (4.31) shows that the N^2 quantities e^{kq} (k, q = 1, 2,..., N) are the elements of the inverse of the matrix whose elements are e_{qs}. In particular, the matrices $[e^{kq}]$ and $[e_{qs}]$ are non-singular. This remark is a proof of Theorem 4.26 also by using Theorem 9.5. It is possible to establish from (4.30) and (4.31) that

$$e_s = \sum_{k=1}^{N} e_{sk} e^k \tag{4.33}$$

To illustrate the construction of a reciprocal basis by an algebraic method, consider the real vector space R^2, which we shall represent by the Euclidean plane. Let a basis be given for R^2, which consists of two vectors 45° degrees apart, the first one, e_1, two units long and the second one, e_2, one unit long. These two vectors are illustrated in the following figure.

To construct the reciprocal basis, we first note that from the given information and equation (4.14) we can write

$$e_{11} = 4,\ e_{12} = e_{21} = \sqrt{2},\ e_{22} = 1 \tag{4.34}$$

Writing equations (4.31) out explicitly for the case $N = 2$, we have

$$\begin{aligned} e^{11}e_{11} + e^{12}e_{21} &= 1, & e^{21}e_{11} + e^{22}e_{21} &= 0 \\ e^{11}e_{12} + e^{12}e_{22} &= 0, & e^{21}e_{12} + e^{22}e_{22} &= 1 \end{aligned} \tag{4.35}$$

Substituting (4.34) into (4.35), we find that

$$e^{11} = \frac{1}{2},\ e^{12} = e^{21} = -1/\sqrt{2},\ e^{22} = 2 \tag{4.36}$$

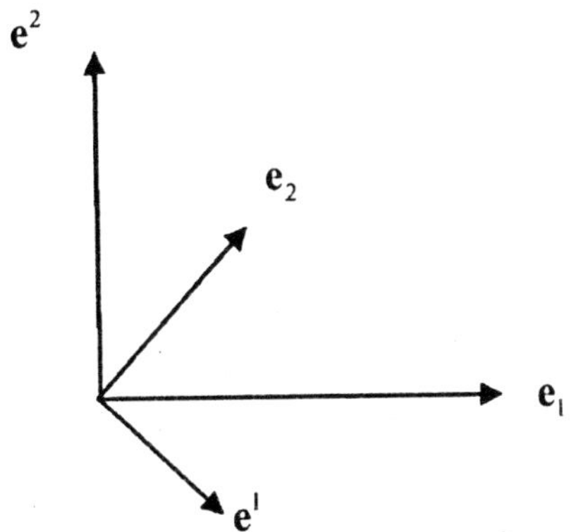

Figure: A basis and reciprocal basis for R^2

When the results (4.26) are put into the special case of (4.30) for $N = 2$, we obtain the explicit expressions for the reciprocal basis $\{e^1, e^2\}$,

$$e^1 = \frac{1}{2}e_1 - \frac{1}{\sqrt{2}}e_2,\ e^2 = -\frac{1}{\sqrt{2}}e_1 + 2e_2 \tag{4.37}$$

The reciprocal basis is illustrated in the figure above also.

Henceforth, we shall use the same kernel letter to denote the components of a vector relative to a basis as we use for the vector itself. Thus, a vector v has components v^k, $k = 1,2,..., N$ relative to the basis $\{e_1, e_2,...,e_N\}$ and components v_k, $k = 1,2,..., N$, relative to its reciprocal basis,

$$v = \sum_{k=1}^{N} v^k e_k, \qquad v = \sum_{k=1}^{N} v_k e^k \tag{4.38}$$

The components $v^1, v^2,..., v^N$ with respect to the basis $\{e_1, e2,..., e_N\}$ are often called the contravariant components of v, while the components, $v_1, v_2,...,v_N$ with respect to the reciprocal basis $\{e^1, e^2,..., e^N\}$ are called covariant components. The names covariant and contravariant are somewhat arbitrary since the basis and reciprocal basis are both bases and we have no particular procedure to choose one over the other. The following theorem illustrates further the same remark.

***Theorem** 4.35:* If $\{e^1,..., e^N\}$ is the reciprocal basis of $\{e_1,...,e_N\}$ then $\{e_1,...,e_N\}$ is also the reciprocal basis of $\{e^1,...,e^N\}$.

For this reason we simply say that the bases $\{e_1,..., e_N\}$ and $\{e^1,...,e^N\}$ are (mutually) reciprocal. The contravariant and covariant components of v are related to one another by the formulas

$$\begin{aligned} v^k &= v \cdot e^k = \sum_{q=1}^{N} e^{qk} v_q \\ v_k &= v \cdot e_k = \sum_{q=1}^{N} e_{qk} v^q \end{aligned} \tag{4.39}$$

where equations (4.14) and (4.32) have been employed. More generally, if u has contravariant components u^i and covariant components u_i relative to $\{e_1,...,e_N\}$ and $\{e^1,...,e^N\}$, respectively, then the inner product of u and v can be computed by the formulas

$$u \cdot v = \sum_{i=1}^{N} u_i \overline{v}^i = \sum_{i=1}^{N} u^i \overline{v_i} = \sum_{i=1}^{N} \sum_{j=1}^{N} e^{ij} u_i \overline{v_j} = \sum_{i=1}^{N} \sum_{j=1}^{N} e_{ij} u^i \overline{v}^j \tag{4.40}$$

which generalise the formulas (4.22) and (4.16).

As an example of the covariant and contravariant components of a vector, consider a vector v in R^2 which has the representation

$$v = \frac{3}{2} e_1 + 2e_2$$

relative to the basis $\{e_1, e_2\}$ illustrated in the previous figure. The contravariant components of v are then (3/2, 2); to compute the covariant components, the formula (4.40) is written out for the case $N = 2$,

$$v_1 = e_{11}\, v^1 + e_{21}\, v^2, \quad v_2 = e_{12}\, v^1 + e_{22}\, v^2$$

Then from (4.34) and the values of the contravariant components the covariant components are given by

$$v_1 = 6 + 2\sqrt{2}, \qquad v_2 = \left(3/\sqrt{2}\right) + 2$$

hence

$$v = \left(6 + 2\sqrt{2}\right)e^1 + \left(3/\sqrt{2} + 2\right)e^2$$

The contravariant components of the vector v are illustrated in the following figure. To develop a geometric feeling for covariant and contravariant components, it is helpful for one to perform a vector decomposition of this type for oneself.

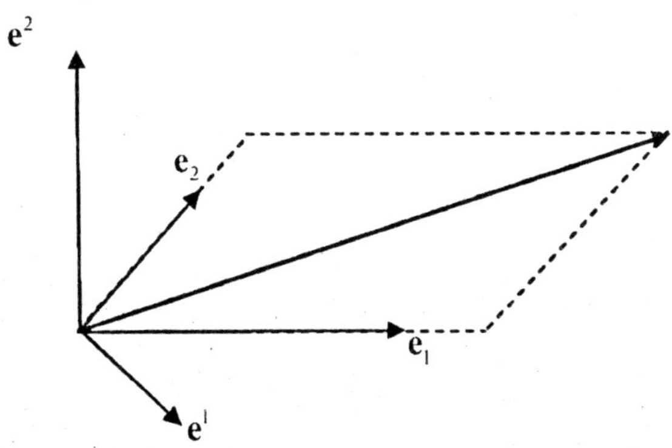

Figure: The covariant and contravariant components of a vector in R^2.

A substantial part of the usefulness of orthonormal bases is that each orthonormal basis is self-reciprocal; hence contravariant and covariant components of vectors coincide. The self-reciprocity of orthonormal bases follows by comparing the condition (4.19) for an orthonormal basis with the condition (4.25) for a reciprocal basis. In orthonormal systems indices are written only as subscripts because there is no need to distinguish between covariant and contravariant components.

Formulas for transferring from one basis to another basis in an inner product space can be developed for both base vectors and components. In these formulas one basis is the set of linearly independent vectors $\{e_1,\ldots, e_N\}$, while the second basis is the set $\{\hat{e}_1,\ldots,\hat{e}_N\}$. From the fact that both bases generate V, we can write

$$\hat{e}_k = \sum_{s=l}^{N} T_k^s e_s, \qquad k = 1,2,\ldots,N \tag{4.41}$$

and

$$e_q = \sum_{k=1}^{N} \hat{T}_q^k \hat{e}_k, \qquad q = 1,2,\ldots,N \tag{4.42}$$

where $\hat{T}_q^k$ and T_k^s are both sets of N^2 scalars that are related to one another. From Theorem 9.5 the $N \times N$ matrices $\left[T_k^s\right]$ and $\left[\hat{T}_q^k\right]$ must be non-singular.

Substituting (4.41) into (4.42) and replacing e_q by $\sum_{s=1}^{N} \delta_q^s e_s$, we obtain

$$e_q = \sum_{k=1}^{N} \sum_{s=1}^{N} \hat{T}_q^k T_k^s e_s = \sum_{s=1}^{N} \delta_q^s e_s$$

which can be rewritten as

$$\sum_{s=1}^{N} \left(\sum_{k=1}^{N} \hat{T}_q^k T_k^s - \delta_q^s \right) e_s = 0 \tag{4.43}$$

The linear independence of the basis $\{e_s\}$ requires that

$$\sum_{k=1}^{N} \hat{T}_q^k T_k^s = \delta_q^s \tag{4.44}$$

A similar argument yields

$$\sum_{k=1}^{N} \hat{T}_k^q T_s^k = \delta_s^q \tag{4.45}$$

In matrix language we see that (4.44) or (4.45) requires that the matrix of elements T_q^k be the inverse of the matrix of elements $\hat{T}_k^q$. It is easy to verify that the reciprocal bases are related by

$$e^k = \sum_{q=1}^{N} \overline{T_q^k} \hat{e}^q, \quad \hat{e}^q = \sum_{k=1}^{N} \overline{\hat{T}_k^q e^k} \tag{4.46}$$

The covariant and contravariant components of $v \in V$ relative to the two pairs of reciprocal bases are given by

$$v = \sum_{k=1}^{N} v_k e^k = \sum_{k=1}^{N} v^k e_k = \sum_{q=1}^{N} \hat{v}^q \hat{e}_q = \sum_{q=1}^{N} \hat{v}_q \hat{e}^q \tag{4.47}$$

To obtain a relationship between, say, covariant components of v, one can substitute $(4.46)_1$ into (4.47), thus

$$\sum_{k=1}^{N} \sum_{q=1}^{N} v_k \overline{T_q^k} \hat{e}^q = \sum_{q=1}^{N} \hat{v}_q \hat{e}^q$$

This equation can be rewritten in the form

$$\sum_{q=1}^{N} \left(\sum_{k=1}^{N} \overline{T_q^k} v_k - \hat{v}_q \right) \hat{e}^q = 0$$

and it follows from the linear independence of the basis $\{\hat{e}^q\}$ that

$$\hat{v}_q = \sum_{k=1}^{N} \overline{T_q^k v_k} \tag{4.48}$$

In a similar manner the following formulas can be obtained:

$$\hat{v}^q = \sum_{k=1}^{N} \hat{T}_k^q v^k \tag{4.49}$$

$$v^k = \sum_{q=1}^{N} T_q^k \hat{v}^q \tag{4.50}$$

and

$$v_k = \sum_{q=1}^{N} \overline{\hat{T}_k^q} \hat{v}_q \tag{4.51}$$

Exercises

1. Show that the quantities T_q^s and $\hat{T}_q^k$ introduced in formulas (4.41) and (4.42) are given by the expressions

   $$T_q^s = \hat{e}_k \cdot e^s, \quad \hat{T}_q^k = e_q \cdot \hat{e}^k \tag{4.52}$$

2. Derive equation (4.45).
3. Derive equations (4.49)-(4.51).
4. Given the change of basis in a three-dimensional inner product space V,

   $$f_1 = 2e_1 - e_2 - e_3, \quad f_2 = -e_1 + e_3, \quad f_3 = 4e_1 - e_2 + 6e_3$$

   find the quantities T_k^s and $\hat{T}_q^k$.
5. Given the vector $v = e_1 + e_2 + e_3$ in the vector space of the problem above, find the covariant and contravariant components of v relative to the basis $\{f_1, f_2, f_3\}$.
6. Show that under the basis transformation (4.41)

   $$\hat{e}_{kj} = \hat{e}_k \cdot \hat{e}_j = \sum_{s,q=1}^{N} T_k^s \overline{T_j^q} e_{sq}$$

   and

   $$\hat{e}^{kj} = \hat{e}^k \cdot \hat{e}^j = \sum_{s,q=1}^{N} \hat{T}_s^k \overline{\hat{T}_q^j} e^{sq}$$

7. Prove Theorem 4.35.

Linear Transformations

Definition of Linear Transformation

In this section and in the other sections of this chapter we shall introduce and study a special class of functions defined on a vector space. If V and U are vector spaces, a *linear transformation* is a function $A : V \to U$ such that

(a) $A(u + v) = A(u) + A(v)$

(b) $A(\lambda u) = \lambda A(u)$

for all $u, v \in V$ and $\lambda \in C$. Condition (a) asserts that A is a *homomorphism* on V with respect to the operation of addition. Condition (b) shows that A, in addition to being a homomorphism, is also *homogeneous* with respect to the operation of scalar multiplication. Observe that the + symbol on the left side of (a) denotes addition in V, while on the right side it denotes addition in U. It would be extremely cumbersome to adopt different symbols for these quantities. Further, it is customary to omit the parentheses and write simply Au for $A(u)$ when A is a linear transformation.

Theorem 5.1: If $A : V \to U$ is a function from a vector space V to a vector space U, then, A is a linear transformation if and only if

$$A(\lambda u + \mu v) = \lambda A(u) + \mu A(v)$$

for all $u, v \in V$ and $\lambda, \mu \in C$.

The proof of this theorem is an elementary application of the definition. It is also possible to show that

$$A\left(\lambda^1 v_1 + \lambda^2 v_2 + \cdots + \lambda^R v_R\right) = \lambda^1 A v_1 + \lambda^2 A v_1 + \cdots + \lambda^R A v_R \quad (5.1)$$

for all $v_1, \ldots, v_R \in V$ and $\lambda^1, \ldots, \lambda^R \in C$.

We see that for a linear transformation $A : V \to U$

$$A0 = 0 \text{ and } A(-v) = -Av \quad (5.2)$$

Note that in (5.2) we have used the same symbol for the zero vector in V as in U.

Theorem 5.2: If $\{v_1, v_2, \ldots, v_R\}$ is a linearly dependent set in V and if $A : V \to U$ is a linear transformation, then $\{Av_1, Av_2, \ldots, Av_R\}$ is a linearly dependent set in U.

Proof: Since the vectors $v_1, \ldots, v_R$ are linearly dependent, we can write

$$\sum_{j=1}^{R} \lambda^j v_j = 0$$

where at least one coefficient is not zero. Therefore,

$$A\left(\sum_{j=1}^{R} \lambda^j v_j\right) = \sum_{j=1}^{R} \lambda^j A v_j = 0$$

where (5.1) and (5.2) have been used. The last equation proves the theorem.

If the vectors $v_1, \ldots, v_R$ are linearly independent, then their image set $\{Av_1, Av_2, \ldots, Av_R\}$ may or may not be linearly independent. For example, A might map all vectors into 0. The *kernel* of a linear transformation $A : V \to U$ is the set

$$K(A) = \{v \mid Av = 0\}$$

In other words, $K(A)$ is the preimage of the set $\{0\}$ in U. Since $\{0\}$ is a subgroup of the additive group U, $K(A)$ is a subgroup of the additive group V. However, a stronger statement can be made.

Theorem 5.3: The set $K(A)$ is a subspace of V.

Proof: Since $K(A)$ is a subgroup, we only need to prove that if $v \in K(A)$, then $\lambda v \in K(A)$ for all $\lambda \in C$. This is clear since

$$A(\lambda v) = \lambda Av = 0$$

for all v in $K(A)$.

The kernel of a linear transformation is sometimes called the *null space*. The nullity of a linear transformation is the dimension of the kernel, i.e., dim $K(A)$. Since $K(A)$ is a subspace of V, we have,

$$\dim k(A) \le \dim V \tag{5.3}$$

Theorem 5.4: A linear transformation $A : V \to U$ is one-to-one if and only if $K(A) = \{0\}$.

Proof: This theorem is just a special case. For ease of reference, we shall repeat the proof. If $Au = Av$ then by the linearity of A, $A(u - v) = 0$. Thus, if $K(A) = \{0\}$, then $Au = Av$ implies $u = v$, so that A is one-to-one. Now assume A is one-to-one. Since, $K(A)$ is a subspace, it must contain the zero in V and therefore $A0 = 0$. If $K(A)$ contained any other element v we would have, $Av = 0$ which contradicts the fact that A is one-to-one.

Linear transformations that are one-to-one are called regular *linear transformations*. The following theorem gives another condition for such linear transformations.

Theorem 5.5: A linear transformation $A : V \to U$ is regular if and only if it maps linearly independent sets in V to linearly independent sets in U.

Proof: Let $\{v_1, v_2, \ldots, v_R\}$ be a linearly independent set in V and $A : V \to U$ be a regular linear transformation. Consider the sum

$$\sum_{j=1}^{R} \lambda^j Av_j = 0$$

This equation is equivalent to

$$A\left(\sum_{j=1}^{R} \lambda^j v_j\right) = 0$$

Since A is regular, we must have

$$\sum_{j=1}^{R} \lambda^j v_j = 0$$

Since the vectors $v_1, v_2, \ldots, v_R$ are linearly independent, this equation shows $\lambda^1 = \lambda^2 = \cdots = \lambda^R = 0$, which implies that the set $\{Av_1, \ldots, Av_R\}$ is linearly independent.

Sufficiency: The assumption that A preserves linear independence implies, in particular, that $Av \neq 0$ for every non-zero vector $v \in V$ since such a vector forms a linearly independent set.

Therefore, $K(A)$ consists of the zero vector only, and thus A is regular.

For a linear transformation $A : V \to U$ we denote the range of A by

$$\mathrm{R(A)} = \{Av \mid v \in V\}$$

It follows that $R(A)$ is a subgroup of U.

Theorem 5.6: The range $R(A)$ is a subspace of U.

We have from Theorems 5.6 that

$$\dim R(A) \leq \dim U \tag{5.4}$$

The rank of a linear transformation is defined as the dimension of $R(A)$, i.e., dim $RA(A)$. A stronger statement than (5.4) can be made regarding the rank of linear transformation A.

Theorem 5.7: $\dim R(A) \leq \min(\dim V, \dim U)$.

Proof: Clearly, it suffices to prove that $\dim R(A) \leq \dim V$, since this inequality and (5.4) imply the assertion of the theorem. Let $\{e_1, e_2, \ldots, e_N\}$ be a basis for V, where $N = \dim V$.

Then, $v \in V$ can be written

$$v = \sum_{j=1}^{N} v^j e^j$$

Therefore, any vector $Av \in R(A)$ can be written

$$Av = \sum_{j=1}^{N} v^j Ae_j$$

Hence, the vectors $\{Ae_1, Ae_2, \ldots, Ae_N\}$ generate R(A). By Theorem 9.10, we can conclude

$$\dim R(A) \leq \dim V \tag{5.5}$$

A result that improves on (5.5) is the following important theorem.

***Theorem** 5.8:* If $A: V \rightarrow U$ is a linear transformation, then

$$\dim V = \dim R(A) + \dim K(A) \tag{5.6}$$

Proof: Let $P = \dim K(A)$, $R = \dim R(A)$, and $N = \dim V$. We must prove that $N = P + R$.

Select N vectors in V such that $\{e_1, e_2, \ldots e_p\}$ is a basis for $k(A)$ and $\{e_1, e_2, \ldots e_p, e_{p+1}, \ldots, e_N\}$ a basis for V. As in the proof of Theorem 5.7, the vectors $\{Ae_1, Ae_2, \ldots Ae_p, Ae_{p+1}, \ldots, Ae_N\}$, generate R(A). But, by the properties of the kernel, $Ae_1 = Ae_1 = \ldots \; Ae_p = 0$. Thus the vectors $\{Ae_{p+1}, Ae_{P+2}, \ldots, Ae_N\}$ generate $R(A)$. If we can establish that these vectors are linearly independent, we can conclude that the vectors $\{Ae_{p+1}, Ae_{P+2}, \ldots, Ae_N\}$ form a basis for R(A) and that $\dim R(A) = R = N - P$.

Consider the sum

$$\sum_{j=1}^{R} \lambda^j Ae_{p+j} = 0$$

Therefore,

$$A\left(\sum_{j=1}^{R} \lambda^j e_{p+j}\right) = 0$$

which implies that the vector $\sum_{j=1}^{R} \lambda^j e_{p+j} \in K(A)$. This fact requires that $\lambda^1 = \lambda^2 = \cdots = \lambda^R = 0$, or otherwise the vector $\sum_{j=1}^{R} \lambda^j e_{p+j}$ could be expanded in the basis $\{e_1, e_2, \ldots, e_p\}$, contradicting the linear independence of $\{e_1, e_2, \ldots, e_N\}$. Thus, the set $\{Ae_{p+1}, Ae_{p+2}, \ldots, Ae_N\}$ is linearly independent and the proof of the theorem is complete.

As usual, a linear transformation $A: V \to U$ is said to be *onto* if $R(A)=U$, i.e., for every vector $u \in U$ there exists $v \in V$ such that $Av = u$.

Theorem 5.9: $\dim R(A) = \dim U$ if and only if A is onto.

In the special case when $\dim V = \dim U$, it is possible to state the following important theorem.

Theorem 5.10: If $A: V \to U$ is a linear transformation and if $V = \dim U$, then A is a linear, transformation onto U if and only if A is regular.

Proof: Assume that $A: V \to U$ is onto u, then (5.6) and Theorem 5.9 show that

$$\dim V = \dim U = \dim U + \dim K(A)$$

Therefore, dim $K(A) = \{0\}$ and thus $K(A) = \{0\}$ and A is one-to-one. Next assume that A is one-to-one. By Theorem 5.4, $K(A) = \{0\}$ and thus dim $K(A) = 0$. Then (5.6) shows that

$$\dim V = \dim R(A) = \dim U$$

We can conclude that

$$R(A) = U$$

and thus A is onto.

Exercises

1. Prove Theorems 5.1, 5.6, and 5.9
2. Let $A : V \to U$ be a linear transformation, and let V^1 be a subspace of V. The restriction of A to V_1 is a function $A|_{V^1}: V^1 \to U$ defined by

$$A|_{V^1}\, v = Av$$

for all $v \in V^1$. Show that $A|_{V^1}$ is a linear transformation and that

$$K(A|_{V^1}) = K(A) \cap V^1$$

3. Let $A : V \to U$ be a linear transformation, and define a function

   $\bar{A} : V/K(A) \to U$ by

   $\bar{A}\bar{v} = Av$

   for all $v \in V$. Here, $\bar{v}$ denotes the equivalence set of v in $V/K(A)$. Show that $\bar{A}$ is a linear transformation. Show also that $\bar{A}$ is regular and that $R(A) = R(\bar{A})$.

4. Let V^1 be a subspace of V, and define a function $P : V \to V/V^1$ by

   $$Pv = \bar{v}$$

   for all v in V. Prove that P is a linear transformation, onto, and that $K(P)=V^1$. The mapping P is called the *canonical projection* from V to V/V^1.

5. Prove the formula

   $$\dim V = \dim(V/V^1) + \dim V^1 \qquad (5.7)$$

   by applying Theorem 5.8 to the canonical projection P defined in the preceding exercise. Conversely, prove the formula (5.6) of Theorem 5.8 by using the formula (5.7).

6. Let U and V be vector spaces, and let $U \oplus V$ be their direct sum. Define mapping $P_1 : U \oplus V \to U$ and $P_2 : U \oplus V \to V$ by

   $$P_1(u, v) = u, \quad P_2(u, v) = v$$

   for all $u \in U$ and $v \in V$. Show that P_1 and P_2 are onto linear transformations. Show also the formula

   $$\dim(U \oplus V) = \dim U + \dim V$$

   by using these linear transformations and the formula (5.6). The mappings P_1 and P_2 are also called the *canonical projections* from $U \oplus V$ to U and V, respectively.

Sums and Products of Liner Transformations

In this section we shall assign meaning to the operations of addition and scalar multiplication for linear transformations. If A

and B are linear transformations $V \to U$, then their sum $A + B$ is a linear transformation defined by

$$(A+B)v = Av+Bv \tag{5.8}$$

for all $v \in V$. In a similar fashion, if $\lambda \in C$, then λA is a linear transformation, $V \to U$ defined by

$$(\lambda A)v = \lambda(Av) \tag{5.9}$$

for all $v \in V$. If we write $L(V;U)$ for the set of linear transformations from V to U, then (5.8) and (5.9) make $L(V;U)$ a *vector space*. The zero element in $L(V;U)$ is the linear transformation 0 defined by

$$0v = 0 \tag{5.10}$$

for all $v \in V$. The negative of $A \in L(V;U)$ is a linear transformation $-A \in L(V;U)$ defined by

$$-A = -1A \tag{5.11}$$

It follows from (5.11) that $-A$ is the additive inverse of $A \in L(V;U)$. This assertion follows from

$$A+(-A) = A+(-1A) = 1A+(-1A)(1-1)A = 0A = 0 \tag{5.12}$$

where (5.8) and (5.9) have been used. Consistent with our previous notation, we shall write $A - B$ for the sum $A + (-B)$ formed from the linear transformations A and B. The formal proof that $L(V;U)$ is a vector space is left as an exercise to the reader.

Theorem 5.11: $\dim L(V;U) = \dim V \dim U$.

Proof: Let $\{e_1,\ldots,e_N\}$ be a basis for V and $\{b_1,\ldots,b_M\}$ be a basis for U. Define NM linear transformations $A_a^k : V \to U$ by

$$A_\alpha^k e_k = b_\alpha,\ k = 1,\ldots,N;\ \alpha = 1,\ldots,M \tag{5.13}$$

$$A_\alpha^k e_p = 0,\ k \neq p$$

If A is an arbitrary member of $L(V,U)$, then $Ae_k \in U$, and thus

$$Ae_k = \sum_{\alpha=1}^{M} A_k^\alpha b_\alpha,\ k = 1,\ldots,N$$

Based upon the properties of A_α^k, we can write the above equation as

$$Ae_k = \sum_{\alpha=1}^{M} A_k^\alpha A_\alpha^k e_k = \sum_{\alpha=1}^{M}\sum_{s=1}^{N} A_s^\alpha A_\alpha^s e_k$$

Therefore, since the vectors $e_1,\ldots,e_N$ generate V, we find

$$\left(A - \sum_{\alpha=1}^{M}\sum_{s=1}^{N} A_s^\alpha A_\alpha^s\right) v = 0$$

for all vectors $v \in V$. Thus, from (5.14),

$$A = \sum_{\alpha=1}^{M}\sum_{s=1}^{N} A_s^\alpha A_\alpha^s$$

This equation means that the MN linear transformations A_α^s $(s = 1,\ldots,N;\ \alpha = 1,\ldots,M)$ generate $L(V;U)$. If we can prove that these linear transformations are linearly independent, then the proof of the theorem is complete. To this end, set

$$\sum_{\alpha=1}^{M}\sum_{s=1}^{N} A_s^\alpha A_\alpha^s = 0$$

Then, from (5.13),

$$\sum_{\alpha=1}^{M}\sum_{s=1}^{N} A_s^\alpha A_\alpha^s (e_p) = \sum_{\alpha=1}^{M} A_p^\alpha b_\alpha = 0$$

Thus $A_p^\alpha = 0$, $(p = 1,\ldots,N;\ \alpha = 1,\ldots,M)$ because the vectors $b_1,\ldots,b_M$ are linearly independent in U. Hence, $\{A_\alpha^s\}$ is a basis of $L(V;U)$. As a result, we have

$$\dim L(V;U) = MN = \dim U \dim V \tag{5.15}$$

If $A : V \to U$ and $B : U \to W$ are linear transformations, their product is a linear transformation $V \to W$, written BA, defined by

$$BAv = B(Av) \tag{5.16}$$

for all $v \in V$. The properties of the product operation are summarised in the following theorem.

Theorem 5.12:

$$\begin{aligned} C(BA) &= (CB)A \\ (\lambda A+\mu B)C &= \lambda AC+\mu BC \\ C(\lambda A+\mu B) &= \lambda CA+\mu CB \end{aligned} \tag{5.17}$$

for all $\lambda, \mu \in C$ and where it is understood that, A, B and C are defined on the proper vector spaces so as to make the indicated products defined.

The proof of Theorem 5.12 is left as an exercise to the reader.

Exercises

1. Prove that $L(V;U)$ is a vector space.
2. Prove Theorem 5.12.
3. Let V, U and W be vector spaces. Given any linear mappings $A: V \to U$ and $B: U \to W$, show that
$$\dim R(BA) \leq \min\,[\dim R(A), \dim R(B)]$$
4. Let $A: V \to U$ be a linear transformation and define $\bar{A}: V/K(A) \to U$. If $P: V \to V/K(A)$ is the canonical projection, show that
$$A = \bar{A}P$$

This result show that every linear transformation can be written as the composition of an onto linear transformation and a regular linear transformation.

Special Types of Linear Transformations

In this section we shall examine the properties of several special types of linear transformations. The first of these is one called an *isomorphism*. A vector space isomorphism is a regular onto linear transformation $A: V \to U$. It immediately follows from Theorem 5.8 that if $A: V \to U$ is an isomorphism, then

$$\dim V = \dim U \tag{5.18}$$

An isomorphism $A: V \to U$ establishes a one-to-one correspondence between the elements of V and U. Thus, there exists a unique inverse function $B: U \to V$ with the property that if

$$u = Av \tag{5.19}$$

then

$$v = B(u) \tag{5.20}$$

for all $u \in U$ and $v \in V$. We shall now show that B is a linear transformation. Consider the vectors u_1 and $u_2 \in U$ and the corresponding vectors [as a result of (5.19) and (5.20)] v_1 and $v_2 \in V$. Then by (5.19), (5.20), and the properties of the linear transformation A,

$$\begin{aligned} B(\lambda u_1 + \mu u_2) &= B(\lambda A v_1 + \mu A v_2) \\ &= B(A(\lambda v_1 + \mu v_2)) \\ &= \lambda v_1 + \mu v_2 \\ &= \lambda B(u_1) + \mu B(u_2) \end{aligned}$$

Thus, B is a linear transformation. This linear transformation shall be written A^{-1}. Clearly the linear transformation A^{-1} is also an isomorphism whose inverse is A; i.e.,

$$(A^{-1})^{-1} = A \tag{5.21}$$

Theorem 5.13: If $A: V \to U$ and $B: U \to W$ are isomorphisms, then $BA: V \to W$ is an isomorphism whose inverse is computed by

$$(BA)^{-1} = A^{-1}B^{-1} \tag{5.22}$$

Proof: The fact that BA is an isomorphism follows directly from the corresponding properties of A and B. The fact that the inverse of BA is computed by (5.22) follows directly because if

$$u = Av \text{ and } w = Bu$$

then

$$v = A^{-1}u \text{ and } u = B^{-1}w$$

Thus,

$$v = (BA)^{-1}w = A^{-1}B^{-1}w$$

Therefore, $\left((BA)^{-1} - A^{-1}B^{-1}\right)w = 0$ for all $w \in W$, which implies (5.22).

The identity linear transformation $I : V \to V$ is defined by

$$Iv = v \tag{5.23}$$

for all v in V. Often it is desirable to distinguish the identity linear transformations on different vector spaces. In these cases we shall denote the identity linear transformation by I_V. It follows from (5.18) and (5.20) that if A is an isomorphism, then

$$AA^{-1} = I_U \quad \text{and} \quad A^{-1}A = I_V \tag{5.24}$$

Conversely, if A is a linear transformation from V to U, and if there exists a linear transformation $B : U \to V$ such that $AB = I_U$ and $BA = I_V$, then A is an isomorphism and $B = A^{-1}$. The proof of this assertion is left as an exercise to the reader. Isomorphisms are often referred to as *invertible* or *non-singular* linear transformations.

A vector space V and a vector space U are said to be isomorphic if there exists at least one isomorphism from V to U.

Theorem 5.14: Two finite-dimensional vector spaces V and U are isomorphic if and only if they have the same dimension.

Proof: Clearly, if V and U are isomorphic, by virtue of the properties of isomorphisms, dim V = dim U. If U and V have the same dimension, we can construct a regular onto linear transformation $A : V \to U$ as follows. If $\{e_1, \ldots, e_N\}$ is a basis for V and $\{b_1, \ldots, b_N\}$ is a basis for U, define A by

$$Ae_k = b_k, \quad k = 1, \ldots, N \tag{5.25}$$

Or, equivalently, if

$$\text{v} = \sum_{k=1}^{N} v^k e^k$$

then define A by

$$Av = \sum_{k=1}^{N} v^k b_k \tag{5.26}$$

A is regular because if $Av = 0$, then $v = 0$. Theorem 5.10 tells us A is onto and thus is an isomorphism.

As a corollary to Theorem 5.14, we see that V and the vector space C^N, where N = dim$V$, are isomorphic.

We have introduced the notation L(V; U) for the vector space of linear transformations from V to U. The set $L(V;V)$ corresponds to the vector space of linear transformations $V \rightarrow V$. An element of $L(V;V)$ is called an *endomorphism* of V. This nomenclature parallels the previous usage of the word endomorphism. If an endomorphism is regular (and thus onto), it is called an *automorphism*. The identity linear transformation defined by (5.23) is an example of an automorphism. If $A \in L(V;V)$,then it is easily seen that

$$AI = IA = A \tag{5.27}$$

Also, if A and are in $L(V;V)$, it is meaningful to compute the products AB and BA; however,

$$AB \neq BA \tag{5.28}$$

in general. For example, let V be a two-dimensional vector space with basis $\{e_1, e_2\}$ and define A and B by the rules

$$Ae_k = \sum_{j=1}^{2} A_k^j e_j \quad \text{and} \quad Be_k = \sum_{j=1}^{2} B_k^j e_j$$

where A_k^j and B_k^j . k, j = 1, 2, are prescribed. Then

$$BAe_k = \sum_{j,l=1}^{2} A_k^j B_j^l e_l \quad \text{and} \quad ABe_k = \sum_{j,l=1}^{2} B_k^j A_j^l e_l$$

An examination of these formulas shows that it is only for special values of A_k^j and B_k^j that $AB = BA$.

The set $L(V;V)$ has defined on it three operations. They are (a) addition of elements $L(V;V)$, (b) multiplication of an element of $L(V;V)$ by a scalar, and (c) the product of a pair of elements of $L(V;V)$. The operations (a) and (b) make $L(V;V)$ into a vector space, while it is easily shown that the operations (a) and (c) make

$L(V;V)$ into a ring. The structure $L(V;V)$ is an example of an associative algebra.

The subset of $L(V;V)$ that consists of all automorphisms of V is denoted by $GL(V)$. It is immediately apparent that $GL(V)$ is not a subspace of $L(V;V)$, because the sum of two of its elements need not be an automorphism. However, this set is easily shown to be a group with respect to the product operation. This group is called the *general linear group*. Its identity element is I and if $A \in GL(V)$, its inverse is $A^{-1} \in GL(V)$.

A projection is an endomorphism $P \in L(V;V)$ which satisfies the condition

$$P^2 = P \tag{5.29}$$

The following theorem gives an important property of a projection.

Theorem 5.15: If $P : V \to V$ is a projection, then

$$V = R(P) \oplus K(P) \tag{5.30}$$

Proof: Let v be an arbitrary vector in V. Let

$$w = v - Pv \tag{5.31}$$

Then, by (5.29), $Pw = Pv - P(Pv) = Pv - Pv = 0$. Thus, $w \in K(P)$.

Since $Pv \in R(P)$, (5.31) implies that

$$V = R(P) + K(P)$$

To show that $R(P) \cap K(P) = \{0\}$, let $u \in R(P) \cap K(P)$. Then, since $u \in R(P)$ for some $v \in V, u = Pv$. But, since u is also in $K(P)$,

$$0 = Pu = P(Pv) = Pv = u$$

which completes the proof.

Then name projection arises from the geometric interpretation of (5.30). Given any $v \in V$, then there are unique vectors $u \in R(P)$ and $w \in K(P)$ such that

$$v = u + w \tag{5.32}$$

where

$$Pu = u \quad \text{and} \quad Pw = 0 \tag{5.33}$$

Geometrically, P takes v and projects in onto the subspace $R(P)$ along the subspace $K(P)$. Figure given below illustrates this point for $V = \mathrm{R}^2$.

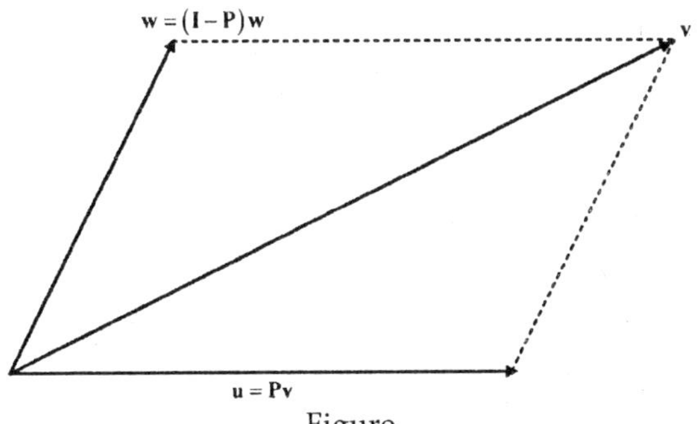

Figure

Given a projection P, the linear transformation I – P is also a projection. It is easily shown that

$$V = R(I-P) \oplus K(I-P)$$

and

$$R(I-P) = K(P), \quad K(I-P) = R(P)$$

It follows from (5.33) that the restriction of P to $R(P)$ is the identity linear transformation on the subspace $R(P)$. Likewise, the restriction of $I - P$ to $K(P)$ is the identity linear transformation on $K(P)$. Theorem 5.15 is a special case of the following theorem.

***Theorem** 5.16:* If $P_k, k = 1, \ldots, R,$ are projection operators with the properties that

$$P_k^2 = P_k, \; k = 1, \ldots, R \tag{5.34}$$

$$P_k P_q = 0, \; k \neq q$$

and

$$I = \sum_{k=1}^{R} P_k \tag{5.35}$$

then

$$V = R(P_1) \oplus R(P_2) \oplus \cdots \oplus R(P_R) \tag{5.36}$$

The proof of this theorem is left as an exercise for the reader. As a converse of Theorem 5.16, if V has the decomposition

$$V = V_1 \oplus \cdots \oplus V_R \tag{5.37}$$

then the endomorphisms $P_k : V \to V$ defined by

$$P_k v = v_k, \quad k = 1, \cdots, R \tag{5.38}$$

where

$$v = v_1 + v_2 + \cdots + v_R \tag{5.39}$$

are projections and satisfy (5.35). Moreover,

$V_k = R(P_k), k = 1, \ldots, R.$

Exercises

1. Let $A : V \to U$ and $B : U \to V$ be linear transformations. If $AB = I$ then B is the right inverse of A. If $BA = I$, then B is the left inverse of A. Show that A is an isomorphism if and only if it has a right inverse and a left inverse.
2. Show that if an endomorphism A of V commutes with every endomorphism B of V, then A is a scalar multiple of I.
3. Prove Theorem 5.16.
4. An involution is an endomorphism L such that $L^2 = I$. Show that L is an involution if and only if $P = \frac{1}{2}(L + I)$ is a projection.
5. Consider the linear transformation $A|_{V^1} : V^1 \to U$. Show that if $V = V^1 \oplus K(A)$, then $A|_{V^1}$ is am isomorphism from V^1 to $R(A)$.
6. If A is a linear transformation from V to U where $\dim V = \dim U$, assume that there, exists a linear transformation $B : U \to V$ such that $AB = I$(or $BA = I$). Show that A is an isomorphism and that $B = A^{-1}$.

Adjoint of a Linear Transformation

In this section, a particular inner product is needed to study the adjoint of a linear transformation as well as other ideas associated with the adjoint.

Given a linear transformation $A: V \to U$, a function $A^*: U \to V$ is called the *adjoint* of A if

$$u \cdot (Av) = (A^* u) \cdot v \tag{5.40}$$

for all $v \in V$ and $u \in U$. Observe that in (5.40) the inner product on the left side is the one in U, while the one for the right side is the one in V. Next we will want to examine the properties of the adjoint. It is probably worthy of note here that for linear transformations defined on real inner product spaces what we have called the *adjoint* is often called the *transpose*. Since our later applications are for real vector spaces, the name transpose is actually more important.

Theorem 5.17: For every linear transformation $A: V \to u$, there exists a unique adjoint $A^*: U \to V$ satisfying the condition (5.17).

Proof: Choose a basis $\{e_1, \ldots, e_N\}$ for V and a basis $\{b_1, \ldots, b_M\}$ for U. Then, A can be characterised by the M × N matrix $\left[A_k^\alpha\right]$ in such a way that

$$Ae_k = \sum_{\alpha=1}^{M} A_k^\alpha b_\alpha, \quad k = 1, \ldots, N \tag{5.41}$$

This system suffices to define A since for any $v \in V$ with the representation

$$v = \sum_{k=1}^{N} v^k e_k$$

the corresponding representation of Av is determined by $\left[A_k^\alpha\right]$ and $\left[v^k\right]$ by

$$Av = \sum_{\alpha=1}^{M} \left(\sum_{k=1}^{N} A_k^\alpha v^k \right) b_\alpha$$

Now let $\{e^1,\ldots,e^N\}$ and $\{b^1,\ldots,b^M\}$ be the reciprocal bases of $\{e_1,\ldots,e_N\}$ and $\{b_1,\ldots,b_M\}$, respectively. We shall define a linear transformation A* by a system similar to (5.41) except that we shall use the reciprocal bases. Thus we put

$$A^* b^\alpha = \sum_{k=1}^{N} A_k^{*\alpha} e^\alpha \tag{5.42}$$

where the matrix $\left[A_k^{*\alpha}\right]$ is defined by

$$A_k^{*\alpha} \equiv \bar{A}_k^a \tag{5.43}$$

for all $\alpha = 1,\ldots, M$ and $k = 1,\ldots,N$. For any vector $u \in U$ the representation of $A^* u$ relative to $\{e^1,\ldots,e^N\}$ is then given by

$$A^* u = \sum_{k=1}^{N}\left(\sum_{\alpha=1}^{M} A_k^{*\alpha} u_\alpha\right) e^k$$

where the representation of u itself relative to $\{b^1,\ldots,b^M\}$ is

$$u = \sum_{\alpha=1}^{M} u_\alpha b^\alpha$$

Having defined the linear transformation A*, we now verify that A and A* satisfy the relation (5.40). Since A and A* are both linear transformations, it suffices to check (5.40) for u equal to an arbitrary element of the basis $\{b^1,\ldots,b^M\}$, say $u = b^\alpha$, and for v equal to an arbitrary element of the basis $\{e_1,\ldots,e_N\}$, say $v = e_k$. For this choice of u and v we obtain from (5.41)

$$b^\alpha \cdot Ae_k = b^\alpha \cdot \sum_{\beta=1}^{M} A_k^\beta b_\beta = \sum_{\beta=1}^{M} \bar{A}_k^\beta \delta_\beta^\alpha = \bar{A}_k^\alpha \tag{5.44}$$

and likewise from (5.42)

$$A^* b^\alpha \cdot e_k = \left(\sum_{l=1}^{N} A_l^{*\alpha} e^l\right) \cdot e_k = \sum_{l=1}^{N} A_l^{*\alpha} \delta_k^l = A_k^{*\alpha} \tag{5.45}$$

Comparing (5.44) and (5.45) with (5.43), we see that the linear transformation A* defined by (5.42) satisfies the condition

$$b^\alpha \cdot Ae_k = A^* b^\alpha \cdot e_k$$

for all $\alpha = 1,\ldots, M$ and $k = 1,\ldots, N$, and hence also the condition (5.40) for all $u \in U$ and $v \in V$.

Uniqueness: Assume that there are two functions $A_1^* : U \to V$ and $A_2^* : U \to V$ which satisfy (5.40). Then

$$\left(A_1^* u\right)\cdot v = \left(A_2^* u\right)\cdot v = u\cdot(Av)$$

Thus,

$$\left(A_1^* u - A_2^* u\right)\cdot v = 0$$

Since the last formula must hold for all $v \in V$, the inner product properties show that

$$A_1^* u = A_2^* u$$

This formula must hold for every $u \in U$ and thus

$$A_1^* = A_2^*$$

As a corollary to the preceding theorem we see that the adjoint A^* of a linear transformation A is a linear transformation. Further, the matrix $\left[A^{*\alpha}_{\ k}\right]$ that characterises A^* by (5.42) is related to the matrix $\left[A_k^\alpha\right]$ that characterises A by (5.41). Notice the choice of bases in (5.41) and (5.42), however.

Other properties of the adjoint are summarised in the following theorem.

Theorem 5.18:

(a) $(A+B)^* = A^* + B^*$ [5.46(a)]

(b) $(AB)^* = B^* A^*$ [5.46(b)]

(c) $(\lambda A)^* = \bar{\lambda} A^*$ [5.46(c)]

(d) $0^* = 0$ [5.46(d)]

(e) $I^* = I$ [5.46(e)]

(f) $(A^*)^* = A$ [5.46(f)]

and

(g) If A is non-singular, so is A^* and in addition,

$$A^{*-1} = A^{-1*} \qquad (5.47)$$

In (a), A and B are in $L(V; U)$; in (b), $B \in L(V;U)$ and $A \in L(U;W)$; in (c), $\lambda \in C$; in(d), 0 is the zero element; and in (e), I is the identity element in $L(V;V)$. The proof of the above theorem is straightforward and is left as an exercise for the reader.

Theorem 5.19: If $A : V \rightarrow U$ is a linear transformation, then V and U have the orthogonal decompositions

$$V = R(A^*) \oplus K(A) \tag{5.48}$$

and

$$U = R(A) \oplus K(A^*) \tag{5.49}$$

where

$$R(A^*) = K(A)^{\perp} \tag{5.50}$$

and

$$R(A) = K(A^*)^{\perp} \tag{5.51}$$

Proof: We shall prove (5.50) and (5.51) and to obtain (5.48) and (5.49). Let u be and arbitrary element in $K(A^*)$. Then for every $v \in V$, $u.(Av) = (A^*u) \cdot v = 0$. Thus $K(A^*)$ is contained in $R(A)^{\perp}$. Conversely, take $u \in R(A)^{\perp}$; then for every, $V \in V (A^*u) \cdot v = u \cdot (Av) = 0$, and thus $(A^*u) = 0$, which implies that $u \in K(A^*)$ and that $R(A)^{\perp}$ is in $K(A^*)$. Therefore, $R(A)^{\perp} = k(A^*)$. Equation (5.50) follows by an identical argument with A replaced by A^*. As mentioned above, (5.48) and (5.49).

Theorem 5.20: Given a linear transformation $A : V \rightarrow U$, then A and A* have the same rank.

Proof: By application of Theorems 5.42,

$$\dim V = \dim R(A^*) + \dim K(A)$$

However, by (5.6),

$$\dim V = \dim R(A) + \dim K(A)$$

Therefore,

$$\dim R(A) = \dim R(A^*) \tag{5.52}$$

which is the desired result.

An endomorphism $A \in L(V;V)$ is called *Hermitian* if A = A* and *skew-Hermitian* if $A = -A^*$. It should be pointed out, however, that the terms *symmetric* and *skew-symmetric* are often used instead of Hermitian and skew-Hermitian for linear transformations defined on real inner product spaces. The following theorem, which follows directly from the above definitions and from (5.17), characterises Hermitian and skew-Hermitian endomorphisms.

***Theorem** 5.21:* An endomorphism A is Hermitian if and only if

$$v_1 \cdot (Av_2) = (Av_1) \cdot v_2 \tag{5.53}$$

for all $v_1, v_2 \in V$, and it is skew-Hermitian if and only if

$$v_1 \cdot (Av_2) = -(Av_1) \cdot v_2 \tag{5.54}$$

for all $v_1, v_2 \in V$.

We shall denote by $S(V;V)$ and $A(V;V)$ the subsets of $L(V;V)$ defined by

$$S(V,V) = \{A \mid A \in L(V;V) \text{and} \, A = A^*\}$$

and

$$A(V;V) = \{A \mid A \in L(V;V) \text{and} \, A = -A^*\}$$

In the special case of a real inner product space, it is easy to show that $S(V,V)$ and $A(V;V)$ are both *subspaces* of $L(V;V)$. In particular, $L(V;V)$ has the following decomposition:

***Theorem** 5.22:* For a real inner product space,

$$L(V;V) = T(V;V) \oplus A(V,V) \tag{5.55}$$

Proof: An arbitrary element $L \in L(V,V)$ can always be written

$$L = S + A \tag{5.56}$$

where

$$S = \frac{1}{2}(L + L^T) \text{ and } A = \frac{1}{2}(L - L^T) = -A^T$$

Here the superscript T denotes that transpose, which is the specialisation of the adjoint for a real inner product space. Since $S \in S(V;V)$ and $A \in A(V;V)$, (5.56) shows that

$$L(V;V) = S(V,V) + A(V;V)$$

Now, let $B \in S(V;V) \cap A(V;V)$. Then B must satisfy the conditions

$$B = B^T \text{ and } B = -B^T$$

Thus, B = 0 and the proof is complete.

A real inner product can be defined on $L(V;V)$ in such a fashion that the subspace of symmetric endomorphisms is the orthogonal complement of the subspace of skew-symmetric endomorphisms. For complex inner product spaces, however, $S(V;V)$ and $A(V;V)$ are not subspaces of $L(V;V)$. For example, if $A \in S(V;V)$, then iA is in $A(V;V)$ because $(iA)^* = \bar{i}A^* = -iA^* = -iA$, where [5.46(c)] has been used.

A linear transformation $A \in L(V;U)$ is unitary if

$$Av_2 \cdot Av_1 = v_2 \cdot v_1 \tag{5.57}$$

for all $v_1 \cdot v_2 \in V$. However, for a real inner product space the above condition defines A to be orthogonal. Essentially, (5.57) asserts that unitary (or orthogonal) linear transformations preserve the inner products.

Theorem 5.23: If A is unitary, then it is regular.

Proof: Take $v_1 = v_2 = v$ in (5.57), we find

$$\|Av\| = \|v\| \tag{5.58}$$

Thus, if $Av = 0$, then $v = 0$, which proves the theorem.

Theorem 5.24: $A \in L(V,U)$ is unitary if and only if $\|Av\| = \|v\|$ for all $v \in V$.

Proof: If A is unitary, we saw in the proof of Theorem 5.23 that $\|Av\| = \|v\|$. Thus, we shall assume $\|Av\| = \|v\|$ for all $v \in V$ and attempt to derive (5.57). This derivation is routine because, by the polar identity,

$$2Av_1 \cdot Av_2 = \|A(v_1+v_2)\|^2 + i\|A(v_1+iv_2)\|^2 - (1+i)\left(\|Av_1\|^2 + \|Av_2\|^2\right)$$

Therefore, by (5 58),

$$2Av_1 \cdot Av_2 = \|(v_1+v_2)\|^2 + i\|A(v_1+iv_2)\|^2 - (1+i)\left(\|v_1\|^2 + \|v_2\|^2\right) = 2v_1 \cdot v_2$$

This proof cannot be specialised directly to a real inner product space since the polar identity is valid for a complex inner product space only. We leave the proof for the real case as an exercise.

If we require V and U to have the same dimension, then Theorem 5.10 ensures that a unitary transformation A is an isomorphism. In the case we can use (5.40) and (5.57) and conclude the following.

***Theorem** 5.25:* Given a linear transformation $A \in L(V;U)$, where $\dim V = \dim U$; then A is unitary if and only if it is an isomorphism whose inverse satisfies

$$A^{-1} = A^* \tag{5.59}$$

Recall from Theorem 5.5 that a regular linear transformation maps linearly independent vectors into linearly independent vectors. Therefore if $\{e_1,\dots,e_N\}$ is a basis for V and $A \in L(V;U)$ is regular, then $\{Ae_1,\dots,Ae_N\}$ is basis for $R(A)$ which is a subspace in U. If $\{e_1,\dots,e_N\}$ is orthonormal and A is unitary, it easily follows that $\{Ae_1,\dots,Ae_N\}$ is also orthonormal. Thus, the image of an orthonormal basis under a unitary transformation is also an orthonormal set. Conversely, a linear transformation which sends an orthonormal basis of V into an orthonormal basis of $R(A)$ must be unitary. To prove this assertion, let $\{e_1,\dots,e_N\}$ be orthonormal, and let $\{b_1,\dots, b_N\}$ be an orthonormal set in U, where, $b_k = Ae_k, k=1,\dots,N$. Then, if v_1 and v_2 are arbitrary elements of V,

$$v_1 = \sum_{k=1}^{N} v_1^k e_k \quad \text{and} \quad v_2 = \sum_{l=1}^{N} v_2^l e_l$$

we have

$$Av_1 = \sum_{k=1}^{N} v_1^l b_k \quad \text{and} \quad Av_2 = \sum_{l=1}^{N} v_2^k b_l\ 1$$

and, thus,

$$Av_1 \cdot Av_2 = \sum_{k=1}^{N}\sum_{l=1}^{N} v_1^k \overline{v}_2 b_k \cdot b_l = \sum_{k=1}^{N} v_1^k \overline{v}_2^l = v_1 \cdot v_2 \qquad (5.60)$$

Equation (5.60) establishes the desired result.

We have introduced the *general linear group* $GL(V)$. We define a subset $U(V)$ of $GL(V)$ by

$$U(V) = \{A \mid A \in GL(V) \text{and}\, A^{-1} = A^*\}$$

which is easily shown to be a subgroup. This subgroup is called the *unitary group of V*.

We have defined projections $P : V \to V$ by the characteristic property

$$P^2 = P \qquad (5.61)$$

In particular, we have showed in Theorem 5.15 that V has the decomposition

$$V = R(P) \oplus K(P) \qquad (5.62)$$

There are several additional properties of projections which are worthy of discussion here.

Theorem 5.26: If P is a projection, then $P = P^* \Leftrightarrow R(P) = K(P)^{\perp}$.

Proof: First take $P = P^*$ and let v be an arbitrary element of V. Then by (5.62)

$$v = u + w$$

where $u = Pv$ and $w = v - Pv$. Then

$$\begin{aligned} u \cdot w &= Pv \cdot (v - Pv) \\ &= (Pv) \cdot v - (Pv) \cdot Pv \\ &= (Pv) \cdot v - (P^* Pv) \cdot v \\ &= (Pv) \cdot v - (P^2 v) \cdot v \\ &= 0 \end{aligned}$$

where (5.61), (5.40), and the assumption that P is Hermitian have been used. Conversely, assume $u.w = 0$ for all $u \in R(P)$ and all $w \in K(P)$. Then if v_1 and v_2 are arbitrary vectors in V

$$Pv_1 \cdot v_2 = Pv_1 \cdot (Pv_2 + v_2 - Pv_2) = Pv_1 \cdot Pv_2$$

and, by interchanging v_1 and v_2

$$Pv_2 \cdot v_1 = Pv_2 \cdot Pv_1$$

Therefore,

$$Pv_1 \cdot v_2 = \overline{Pv_2 \cdot v_1} = v_1 \cdot (Pv_2)$$

This last result and Theorem 5.21 show that P is Hermitian.

Because of Theorem 5.26, Hermitian projections are called *perpendicular projections.* We have introduced the concept of an orthogonal complement of a subspace of an inner product space. A similar concept is that of an orthogonal pair of subspaces. If V_1 and V_2 are subspaces of V, they are orthogonal, written $V_1 \perp V_2$, if $v_1.v_2 = 0$ for all $v_1 \in V_1$ and $v_2 \in V_2$.

Theorem 5.27: If V_1 and V_2 are subspaces of V, P_1 is the perpendicular projection of V onto V_1, and P_2 is the perpendicular projection of P onto v_2, then $V_1 \perp V_2$ if and only if $P_2P_1 = 0$.

Proof: Assume that $V_1 \perp V_2$; then $P_1 v \in V_2^{\perp}$ for all $v \in V$ and thus $P_2P_1v = 0$ for all $v \in V$ which yields $P_2P_1 = 0$. Next assume $P_2P_1 = 0$; this implies that $P_1 v \in V_2^{\perp}$ for every $v \in V$.

Therefore, V_1 is contained in $V_2^{\perp}$ and, as a result, $V_1 \perp V_2$.

Exercises

1. Prove Theorem 5.18.
2. For a real inner product space, prove that $\dim S(V;V) = \frac{1}{2}N(N+1)$ and $\dim A(V;V) = \frac{1}{2}N(N-1)$, where $N = \dim V$.
3. Define $\Phi : L(V;V) \to L(V;V)$ and $\psi : L(V;V) \to L(V;V)$ by

$$\Phi A = \frac{1}{2}(A + A^T) \quad \text{and} \quad \psi A = \frac{1}{2}(A - A^T)$$

where V is a real inner product space. Show that Φ and Ψ are projections.

4. Let V_1 and V_2 be subspaces of V and let P_1 be the perpendicular projection of V onto V_1 and P_2 be the perpendicular projection of V onto V_2. Show that $P_1 + P_2$ is a perpendicular projection if and only if $V_1 \perp V_2$.

5. Let P_1 and P_2 be the projections introduced in Exercise 4. Show that $P_1 - P_2$ is a projection if and only if V_1 is a subspace of V_1.

6. Show that $A \in L(V;V)$ is skew-symmetric $\Leftrightarrow v \cdot Av = 0$ for all $v \in V$, where V is real vector space.

7. A linear transformation $A \in L(V;V)$ is normal if $A^*A = AA^*$. Show that A is normal if and only if

$$Av_1 \cdot Av_2 = A^*v_1 \cdot A^*v_2$$

for all v_1, v_2 in V.

8. Show that $v \cdot ((A + A^*)v)$ is real for every $v \in V$ and every $A \in L(V;V)$.

9. Show that every linear transformation $A \in L(V;V)$, where V is a complex inner product space, has the unique decomposition

$$A = B + iC$$

where B and C are both Hermitian.

10. Prove Theorem 5.24 for real inner product spaces. *Hint:* A polar identity for a real inner product space is

$$2u \cdot v = \|u + v\|^2 - \|u\|^2 - \|v\|^2$$

11. Let A be an endomorphism of R^3 whose matrix relative to the standard basis is

$$\begin{bmatrix} 1 & 2 & 1 \\ 4 & 6 & 3 \\ 1 & 0 & 0 \end{bmatrix}$$

What is $K(A)$? What is $R(A^T)$? Check the results of Theorem 5.19 for this particular endomorphism.

Component Formulas

In this section we shall introduce the components of a linear transformation and several related ideas. Let $A \in L(V;U)$, $\{e_1,\ldots,e_N\}$ a basis for V, and $\{b_1,\ldots,b_M\}$ a basis for U. The vector Ae_k is in U and, as a result, can be expanded in a basis of U in the form

$$Ae_k = \sum_{\alpha=1}^{M} A_k^\alpha b_\alpha \tag{5.63}$$

The MN scalars A_k^α $(\alpha = 1,\ldots,M; k = 1,\ldots,N)$ are called the *components of* A with respect to the bases. If $\{b^1,\ldots,b^M\}$ is a basis of U which is reciprocal to $\{b_1,\ldots,b_M\}$, (5.63) yields

$$A^\alpha{}_k = (Ae_k)\cdot b^\alpha \tag{5.64}$$

Under change of bases in V and U

$$e_k = \sum_{j=1}^{N} \hat{T}_k^j \hat{e}_j \tag{5.65}$$

and

$$b^\alpha = \sum_{\beta=1}^{M} \overline{S_\beta^\alpha} \hat{b}^\beta \tag{5.66}$$

(5.64) can be used to derive the following transformation rule for the components of A:

$$A^\alpha{}_k = \sum_{\beta=1}^{M}\sum_{j=1}^{N} S_\beta^\alpha \hat{A}_j^\beta \hat{T}_k^j \tag{5.67}$$

where

$$\hat{A}_j^\beta = (A\hat{e}_j)\cdot \hat{b}^\beta$$

If $A \in L(V;V)$, (5.63) and (5.67) specialise to

$$Ae_k = \sum_{q=1}^{N} A^q{}_k e_q \tag{5.68}$$

and

$$A^q_k = \sum_{s,j=1}^{N} T^q_s \hat{A}^s_j \hat{T}^j_k \tag{5.69}$$

The trace of an endomorphism is a function tr $: L(V;V) \to C$ defined by

$$trA = \sum_{k=1}^{N} A^k{}_k \tag{5.70}$$

It easily follows from (5.69) that tr A is independent of the choice of basis of V. Later we shall give a definition of the trace which does not employ the use of a basis.

If $A \in L(V;U)$, then $A^* \in L(V;U)$. The components of A^* are obtained by the same logic as was used in obtaining (5.63). For example,

$$A^* b^\alpha = \sum_{k=1}^{N} A^*{}_k{}^\alpha e^K \tag{5.71}$$

where the MN scalars $A^*{}_k{}^\alpha \, (k = 1,\ldots,N; \alpha = 1,\ldots,M)$ are the components of A^* with respect to $\{b^1,\ldots,b^M\}$ and $\{e^1,\ldots,e^N\}$. From the proof of Theorem 5.17 we can relate these components of A^* to those of A in (5.63); namely,

$$A^{*\alpha}_s = \overline{A^\alpha{}_s} \tag{5.72}$$

If the inner product spaces U and V are real, (5.72) reduces to

$$A^T{}_s{}^\alpha = A^\alpha{}_s \tag{5.73}$$

By the same logic which produced (5.63), we can also write

$$Ae_k = \sum_{\alpha=1}^{M} A_{\alpha k} b^\alpha \tag{5.74}$$

and

$$Ae_k = \sum_{\alpha=1}^{M} A^{\alpha k} b_\alpha = \sum_{\alpha=1}^{M} A_\alpha{}^k b^\alpha \tag{5.75}$$

The various components of A are related by

$$A^\alpha{}_k = \sum_{s=1}^{N} A^{\alpha s} e_{ks} = \sum_{\beta=1}^{M}\sum_{s=1}^{N} b^{\beta\alpha} A_\beta{}^s e_{ks} = \sum_{\beta=1}^{M} A_{\beta k} b^{\beta\alpha} \tag{5.76}$$

where

$$b_{\alpha\beta} = b_\alpha \cdot b_\beta = \overline{b_{\beta\alpha}} \text{ and } b^{\alpha\beta} = b^\alpha \cdot b^\beta = \overline{b^{\beta\alpha}} \tag{5.77}$$

A similar set of formulas holds for the components of A^*. Equations (5.76) and (5.72) can be used to obtain these formulas. The transformation rules for the components defined by (5.74) and (5.75) are easily established to be

$$A_{\alpha k} = (Ae_k)\cdot b_\alpha = \sum_{\beta=1}^{M}\sum_{j=1}^{N} \overline{\hat{S}_\alpha^\beta}\, \hat{A}_{\beta j} \hat{T}_k^j \tag{5.78}$$

$$A^{\alpha k} = (Ae^k)\cdot b^\alpha = \sum_{\beta=1}^{M}\sum_{j=1}^{N} S_\beta^\alpha \hat{A}^{\beta j} \overline{T_j^k} \tag{5.79}$$

and

$$A_\alpha{}^k = (Ae^k)\cdot b_\alpha = \sum_{\beta=1}^{M}\sum_{j=1}^{N} \overline{\hat{S}_\alpha^\beta}\, \hat{A}_\beta{}^j \overline{T_j^k} \tag{5.80}$$

where

$$\hat{A}_{\beta j} = (A\hat{e}_j)\cdot \hat{b}_\beta \tag{5.81}$$

$$\hat{A}^{\beta j} = (A\hat{e}^j)\cdot \hat{b}^\beta \tag{5.82}$$

and

$$\hat{A}_\beta{}^j = (A\hat{e}^j)\cdot \hat{b}_\beta \tag{5.83}$$

The quantities $\hat{S}_{\alpha}^{\beta}$ introduced in (5.78) and (5.80) above are related to the quantities S_{β}^{α}.

Exercises

1. Express (5.59), written in the form $A^{*}A = I$, in components.
2. Verify (5.76).
3. Establish the following properties of the trace of endomorphisms of V:

$$\operatorname{tr} I = \dim V$$

$$\operatorname{tr}(A + B) = \operatorname{tr} A + \operatorname{tr} B$$

$$\operatorname{tr}(\lambda A) = \lambda \operatorname{tr} A$$

and $\operatorname{tr} AB = \operatorname{tr} BA$

$$\operatorname{tr} A^{*} = \operatorname{tr} A$$

4. If $A, B \in L(V,V)$, where V is an inner product space, define

   $A \cdot B = \operatorname{tr} AB^{*}$

   Show that this definition makes $L(V; V)$ into an inner product space.
5. In the special case when V is a real inner product space, show that the inner product of Exercise 5.66 implies that $A(V;V) \perp S(V;V)$.
6. Verify formulas (5.78)-(5.80).
7. Given an endomorphism $A \in L(V;V)$, show that

$$\operatorname{tr} A = \sum_{k=1}^{N} A_k^{\ k}$$

   where $A_k^{\ q} = (Ae^q) \cdot e_k$. Thus, we can compute the trace of an endomorphism from (5.70) or from the formula above and be assured the same result is obtained in both cases. Of course, the formula above is also independent of the basis of V.

8. Exercise 5.70 shows that tr A can be computed from two of the four possible sets of components of A. Show that the quantities $\sum_{k=1}^{N} A^{kk}$ and $\sum_{k=1}^{N} A_{kk}$ are not equal to each A. In addition, show that each of these quantities depends upon the choice of basis of V.

9. Show that

$$A^{*k\alpha} = \left(A^{*}b^{\alpha}\right)\cdot e^{k} = \overline{A^{\alpha k}}$$

$$A^{*}{}_{k\alpha} = \left(A^{*}b_{\alpha}\right)\cdot e_{k} = \overline{A_{\alpha k}}$$

and

$$A^{*k}{}_{\alpha} = \left(A^{*}b_{\alpha}\right)\cdot e^{k} = \overline{A_{\alpha}{}^{k}}$$

Determinants and Matrices

In this chapter we shall consider further the concept of a matrix and its relation to a linear transformation.

Kronecker Deltas

Recall that a $M \times N$ matrix is an array written in anyone of the forms

$$A = \begin{bmatrix} A^1{}_1 & \cdot & \cdot & \cdot & A^1{}_N \\ \cdot & & & & \\ \cdot & & & & \\ \cdot & & & & \\ A^M{}_1 & \cdot & \cdot & \cdot & A^M{}_N \end{bmatrix} = \left[A^\alpha{}_j\right], \quad A = \begin{bmatrix} A_{11} & \cdot & \cdot & \cdot & A_{1N} \\ \cdot & & & & \\ \cdot & & & & \\ \cdot & & & & \\ A_{M1} & \cdot & \cdot & \cdot & A_{MN} \end{bmatrix} = \left[A_{\alpha j}\right]$$

$$A = \begin{bmatrix} A_1{}^1 & \cdot & \cdot & \cdot & A_1{}^N \\ \cdot & & & & \\ \cdot & & & & \\ \cdot & & & & \\ A_M{}^1 & \cdot & \cdot & \cdot & A_M{}^N \end{bmatrix} = \left[A_\alpha{}^j\right], \quad A = \begin{bmatrix} A_{11} & \cdot & \cdot & \cdot & A^{1N} \\ \cdot & & & & \\ \cdot & & & & \\ \cdot & & & & \\ A^{M1} & \cdot & \cdot & \cdot & A^{MN} \end{bmatrix} = \left[A^{\alpha j}\right]$$

(6.1)

Throughout this section the placement of the indices is of no consequence. The components of the matrix are allowed to be complex numbers. The set of $M \times N$ matrices shall be denoted by

$M^{M \times N}$. It is an elementary exercise to show that the rules of addition and scalar multiplication insure that $M^{M \times N}$ is a vector space. We have used the set $M^{M \times N}$ as one of several examples of a vector space. We leave as an exercise to the reader the fact that the dimension of $M^{M \times N}$ is MN.

In order to give a definition of a determinant of a square matrix, we need to define a permutation and consider certain of its properties. Consider a set of K elements $\{\alpha_1, ..., \alpha_K\}$. A *permutation* is a one-to-one function from $\{\alpha_1, ..., \alpha_K\}$ to $\{\alpha_1, ..., \alpha_K\}$. If σ is a permutation, it is customary to write

$$\sigma = \begin{pmatrix} \alpha_1 & \alpha_2 & \cdot & \cdot & \cdot & \alpha_K \\ \sigma(\alpha_1) & \sigma(\alpha_2) & \cdot & \cdot & \cdot & \sigma(\alpha_K) \end{pmatrix} \tag{6.2}$$

It is a known result in algebra that permutations can be classified into even and odd ones. The permutation σ in (6.2) is *even* if an even number of pairwise interchanges of the bottom row is required to order the bottom row exactly like the top row, and σ is *odd* if that number is odd. For a given permutation σ, the number of pairwise interchanges required to order the bottom row the same as the top row is not unique, but it is proved in algebra that those numbers are either all even or all odd. Therefore, the definition for σ to be even or odd is meaningful. For example, the permutation

$$\sigma = \begin{pmatrix} 1 & 2 & 3 \\ 2 & 1 & 3 \end{pmatrix}$$

is odd, while the permutation

$$\sigma = \begin{pmatrix} 1 & 2 & 3 \\ 2 & 3 & 1 \end{pmatrix}$$

is even.

The *parity* of σ, denoted by ε_σ, is defined by

$$\varepsilon_\sigma = \begin{cases} +1 \text{ if } \sigma \text{ is an even permutation} \\ -1 \text{ if } \sigma \text{ is an odd permutation} \end{cases}$$

All of the applications of permutations we shall have are for permutations defined on K ($K \le K$) positive integers selected from the set of N positive integers {1,2,3,..., N}. Let $\{i_1,...,i_K\}$ and $\{j_1,...,j_K\}$ be two subsets of {1,2,3,..., N}. If we order these two subsets and construct the two K - tuples $\{i_1,...,i_K\}$ and $\{j_1,...,j_K\}$, we can define the generalised Kronecker delta as follows: The generalised Kronecker delta, denoted by

$$\delta^{i_1 i_2 ... i_k}$$
$$j_1 j_2 \cdots j_k$$

is defined by

$$\delta^{i_1 i_2 \cdots i_K}_{j_1 j_2 \cdots j_K} = \begin{cases} 0 \text{ if the integers } (i_1,...,i_k) \text{ or } (j_1,...,j_k) \text{ are not distinct} \\ 0 \text{ if the integers } (i_1,...,i_K) \text{ and } (j_1,...,j_K) \text{ are distinct} \\ \quad \text{but the sets } (i_1,...,i_K) \text{ and } (j_1,...,j_K) \text{ are not equal} \\ \\ \varepsilon_\sigma \text{ if the integers } (i_1,...,i_K) \text{ and } (j_1,...,j_K) \text{ are distinct} \\ \text{and the sets } (i_1,...,i_K) \text{ and } (j_1,...,j_K) \text{ are equal, where} \\ \sigma - \begin{pmatrix} i_1 & i_2 & \cdot & \cdot & \cdot & i_K \\ j_1 & j_2 & \cdot & \cdot & \cdot & j_K \end{pmatrix} \end{cases}$$

It follows from this definition that the generalised Kronecker delta is zero whenever the superscripts are not the same set of integers as the subscripts, or when the superscripts are not distinct, or when the subscripts are not distinct. Naturally, when $K = 1$ the generalised Kronecker delta reduces to the usual one. As an example, $\delta^{i_1 i_2}_{j_1 j_2}$ has the values

$$\delta^{12}_{12} = 1,\ \delta^{12}_{21} = -1, \delta^{13}_{12} = 0, \delta^{13}_{21} = 0, \delta^{11}_{12} = 1, \text{ etc.}$$

It can be shown that there are $N!K!/(N - K)!$ non-zero generalised Kronecker deltas for given positive integers N and N.

An ε symbol is one of a pair of quantities

$$\varepsilon^{i_1 i_2 ... i_N} = \delta^{i_1 i_2 ... i_N}_{12...N} \text{ or } \varepsilon_{j_1 j_2 ... j_N} = \delta^{12...N}_{j_1 j_2 ... j_N} \tag{6.3}$$

For example, take $N = 3$; then

$$\varepsilon^{123} = \varepsilon^{312} = \varepsilon^{231} = 1$$

$$\varepsilon^{132} = \varepsilon^{321} = \varepsilon^{213} = -1$$

$$\varepsilon^{112} = \varepsilon^{221} = \varepsilon^{222} = \varepsilon^{233} = 0, \text{ etc.}$$

As an exercise, the reader is asked to confirm that

$$\varepsilon^{i_1 \dots i_N} \varepsilon_{j_1 \dots j_N} = \delta^{i_1 \dots i_N}_{j_1 \dots j_N} \tag{6.4}$$

An identity involving the quantity δ^{ijq}_{lms} is

$$\sum_{q=1}^{N} \delta^{ijq}_{lmq} = (N-1)\delta^{ij}_{lm} \tag{6.5}$$

To establish this identity, expand the left side of (6.5) in the form

$$\sum_{q=1}^{N} \delta^{ijq}_{lmq} = \delta^{ij1}_{lm1} + \cdots + \delta^{ijN}_{lmN} \tag{6.6}$$

Clearly we need to verify (6.5) for the cases $i \neq j$ and $\{i, j\} = \{l, m\}$ only, since in the remaining cases (6.5) reduces to the trivial equation $0 = 0$. In the non-trivial cases, exactly two terms on the right-hand side of (6.6) are equal to zero: one for $i = q$ and one for $j = q$. For i, j, l, and m not equal to q, δ^{ijq}_{lmq} has the same value as δ^{ij}_{lm}. Therefore,

$$\sum_{q=1}^{N} \delta^{ijq}_{lmq} = (N-1)\delta^{ij}_{lm}$$

which is the desired result (6.5). By the same procedure used above, it is clear that

$$\sum_{j=1}^{N} \delta^{ij}_{lj} = (N-1)\delta^{i}_{l} \tag{6.7}$$

Combining (6.5)and (6.7), we have

$$\sum_{j=1}^{N}\sum_{q=1}^{N} \delta^{ijq}_{ljq} = (N-2)(N-1)\delta^{i}_{j} \tag{6.8}$$

Since $\sum_{j=1}^{N} \delta^{i}_{i} = N$, we have from (6.8)

$$\sum_{i=1}^{N}\sum_{j=1}^{N}\sum_{q=1}^{N} \delta^{ijq}_{ljq} = (N-2)(N-1)(N) = \frac{N!}{(N-3)!} \tag{6.9}$$

Equation (6.9) is a special case of

$$\sum_{i_1,i_2,\ldots,i_K=1}^{N} \delta^{i_1 i_2 \ldots i_K}_{i_1 i_2 \ldots i_K} = \frac{N!}{(N-K)!} \tag{6.10}$$

Several other numerical relationships are

$$\sum_{i_{R+1},i_{R+2},\ldots,i_K=1}^{N} \delta^{i_1 \ldots i_R i_{R+1} \ldots i_K}_{j_1 \ldots j_R i_{R+1} \ldots i_K} = \frac{(N-R)!}{(N-K)!} \delta^{i_1 \ldots i_R}_{j_1 \ldots j_R} \tag{6.11}$$

$$\sum_{i_{K+1},\ldots,i_N=1}^{N} \varepsilon^{i_1 \ldots i_K i_{K+1} \ldots i_N} \varepsilon_{j_1 \ldots j_K i_{K+1} \ldots i_N} = (N-K)!\,\delta^{i_1 \ldots i_K}_{j_1 \ldots j_K} \tag{6.12}$$

$$\sum_{i_{K+1},\ldots,i_N=1}^{N} \varepsilon^{i_1 \ldots i_K i_{K+1} \ldots i_N} \delta^{j_{K+1} \ldots j_N}_{i_{K+1} \ldots i_N} = (N-K)!\,\varepsilon^{i_1 \ldots i_K j_{K+1} \ldots j_N} \tag{6.13}$$

$$\sum_{j_{K+1},\ldots,j_R=1}^{N} \delta^{i_1 \ldots i_K i_{K+1} \ldots i_R}_{j_i \ldots j_K j_{K+1} \ldots j_R} \delta^{j_{K+1} \ldots j_R}_{l_{K+1} \ldots l_R} = (R-K)!\,\delta^{i_1 \ldots i_K i_{K+1} \ldots i_R}_{j_i \ldots j_K l_{K+1} \ldots l_R} \tag{6.14}$$

$$\sum_{j_{K+1},\ldots,j_R=1}^{N} \sum_{i_{K+1},\ldots,i_R=1}^{N} \delta^{i_1 \ldots i_K i_{K+1} \ldots i_R}_{j_i \ldots j_K j_{K+1} \ldots j_R} \delta^{j_{K+1} \ldots j_R}_{l_{K+1} \ldots i_R} \frac{(N-K)!}{(N-R)!}(R-K)!\,\delta^{i_1 \ldots i_K}_{j_i \ldots j_K} \tag{6.15}$$

It is possible to simplify the formal appearance of many of our equations if we adopt a *summation convention*: We automatically sum every repeated index without writing the summation sign. For example, (6.5) is written

$$\delta^{ijq}_{lmq} = (N-1)\delta^{ij}_{lm} \tag{6.16}$$

The occurrence of the subscript q and the superscript q implies the summation indicated in (6.5). We shall try to arrange things so that we always sum a superscript on a subscript. It is important to know the range of a given summation, so we shall use the summation convention only when the range of summation is understood. Also, observe that the repeated indices are *dummy* indices in the sense that it is unimportant which symbol is used for them. For example,

$$\delta^{ijs}_{lms} = \delta^{ijt}_{lmt} = \delta^{ijq}_{lmq}$$

Naturally there is no meaning to the occurrence of the same index more than twice in a given term.

Other than the summation or dummy indices, many equations have *free* indices whose values in the given range {1,..., N} are arbitrary.

For example, the indices$j$, k, l and m are free indices in (6.16). Notice that every term in a given equation must have the same free indices; otherwise, the equation is meaningless.

The summation convention will be adopted in the remaining portion of this text. If we feel it might be confusing in some context, summations will be indicated by the usual summation sign.

Exercises

1. Verify equation (6.4).
2. Prove that dim $M^{M \times N} = MN$.

Determinants

In this section we shall use the generalised Kronecker deltas and the ε symbols to define the determinant of a square matrix.

The *determinant* of the $N \times N$ matrix $A=[A_{ij}]$, written det A, is a complex number defined by

$$\det A = \begin{vmatrix} A_{11} & A_{12} & \cdot & \cdot & \cdot & A_{1N} \\ A_{21} & A_{22} & \cdot & \cdot & \cdot & A_{2N} \\ A_{31} & & & & & \\ \cdot & & & & & \\ \cdot & & & & & \\ \cdot & & & & & \\ A_{N1} & A_{N2} & \cdot & \cdot & \cdot & A_{NN} \end{vmatrix} = \varepsilon^{i_1 \dots i_N} A_{i_1 1} A_{i_2 2} \dots A_{i_N N} \qquad (6.17)$$

where all summations are from 1 to N. If the elements of the matrix are written $[A^{ij}]$, its determinant is defined to be

$$\det A = \begin{vmatrix} A^{11} & A^{12} & \cdots & A^{1N} \\ A^{21} & A^{22} & \cdots & A^{2N} \\ \cdot & & & \\ \cdot & & & \\ \cdot & & & \\ A^{N1} & A^{N2} & \cdots & A^{NN} \end{vmatrix} = \varepsilon_{i_1 \ldots i_N} A^{i_1 1} A^{i_2 2} \ldots A^{i_N N} \tag{6.18}$$

Likewise, if the elements of the matrix are written $[A^i{}_j]$ its determinant is defined by

$$\det A = \begin{vmatrix} A^1{}_1 & A^1{}_2 & \cdots & A^1{}_N \\ A^2{}_1 & A^2{}_2 & \cdots & A^2{}_N \\ \cdot & & & \\ \cdot & & & \\ \cdot & & & \\ A^N{}_1 & A^N{}_2 & \cdots & A^N{}_N \end{vmatrix} = \varepsilon_{i_1 \ldots i_N} A^{i_1}{}_1 A^{i_2}{}_2 \ldots A^{iN}{}_N \tag{6.19}$$

A similar formula holds when the matrix is written $A=[A_i^j]$. The generalised Kronecker delta can be written as the determinant of ordinary Kronecker deltas. For example, one can show, using (6.19), that

$$\delta^{ij}_{kl} = \begin{vmatrix} \delta^i_k & \delta^i_l \\ \delta^j_k & \delta^j_l \end{vmatrix} = \delta^i_k \delta^j_l - \delta^i_l \delta^j_k \tag{6.20}$$

Equation (6.20) is a special case of the general result

$$\delta^{i_1 \ldots i_K}_{j_1 \ldots j_K} = \begin{vmatrix} \delta^{i_1}_{j_1} & \delta^{i_1}_{j_2} & \cdots & \delta^{i_1}_{j_K} \\ \delta^{i_2}_{j_1} & \delta^{i_2}_{j_2} & \cdots & \delta^{i_2}_{j_K} \\ \cdot & & & \\ \cdot & & & \\ \cdot & & & \\ \delta^{i_K}_{j_1} & \cdots & \cdots & \delta^{i_K}_{j_K} \end{vmatrix} \tag{6.21}$$

It is possible to use (6.17)-(6.19) to show that

$$\varepsilon_{j_1 \dots j_N} \det A = \varepsilon^{i_1 \dots i_N \dots} A_{i_1 j_1} \dots A_{i_N j_N} \tag{6.22}$$

$$\varepsilon^{j_1 \dots j_N} \det A = \varepsilon_{i_1 \dots i_N} A^{i_1 j_1} \dots A^{i_N j_N} \tag{6.23}$$

and

$$\varepsilon_{j_1 \dots j_N} \det A = \varepsilon_{i_1 \dots i_N} A^{i_1}{}_{j_1} \dots A^{i_N}{}_{j_N} \tag{6.24}$$

It is possible to use (6.22)-(6.24) and (6.10) with $N = K$ to show that

$$\det A = \frac{1}{N!} \varepsilon^{j_1 \dots j_N} \varepsilon^{i_1 \dots i_N} A_{i_1 j_1} \dots A_{i_N j_N} \tag{6.25}$$

$$\det A = \frac{1}{N!} \varepsilon_{j_1 \dots j_N} \varepsilon_{i_1 \dots i_N} A^{i_1 j_1 \dots} A^{i_N j_N} \tag{6.26}$$

and

$$\det A = \frac{1}{N!} \delta^{j_1 \dots j_N}_{i_1 \dots i_N} A^{i_1}{}_{j_1} \dots A^{i_N}{}_{j_N} \tag{6.27}$$

Equations (6.25), (6.26) and (6.27) confirm the well-known result that a matrix and its transpose have the same determinant. The reader is cautioned that the transpose mentioned here is that of the matrix and not of a linear transformation. By use of this fact it follows that (6.17)-(6.19) could have been written

$$\det A = \varepsilon^{i_1 \dots i_N} A_{1 i_1} \dots A_{N i_N} \tag{6.28}$$

$$\det A = \varepsilon_{i_1 \dots i_N} A^{1 i_1} \dots A^{N i_N} \tag{6.29}$$

and

$$\det A = \varepsilon^{i_1 \dots i_N} A^1{}_{i_1} \dots A^N{}_{i_N} \tag{6.30}$$

A similar logic also yields

$$\varepsilon_{j_1 \dots j_N} \det A = \varepsilon^{i_1 \dots i_N} A_{j_1 i_1} \dots A_{j_N i_N} \tag{6.31}$$

$$\varepsilon^{j_1 \dots j_N} \det A = \varepsilon_{i_1 \dots i_N} A^{j_1 i_1} \dots A^{j_N i_N} \tag{6.32}$$

and

$$\varepsilon^{j_1 \dots j_N} \det A = \varepsilon^{i_1 \dots i_N} A^{j_1}{}_{i_1} \dots A^{j_N}{}_{i_N} \tag{6.33}$$

The *cofactor* of the element $A^s{}_t$ in the matrix $A=[A^i{}_j]$ is defined by

$$\operatorname{cof} A^s{}_t = \varepsilon_{i_1 i_2 \dots i_N} A^{i_1}{}_1 \dots A^{i_{t-1}}{}_{t-1} \delta^{i_1}_s A^{i_{t+1}}{}_{t+1} \dots A^{i_N}{}_N \tag{6.34}$$

As an illustration of the application of (6.34), let $N=3$ and $s=t=1$. Then

$$\begin{aligned} \operatorname{cof} A^1{}_1 &= \varepsilon_{ijk} \delta^i_1 A^j{}_2 A^k{}_3 \\ &= \varepsilon_{1jk} A^j{}_2 A^k{}_3 \\ &= A^2{}_2 A^3{}_3 - A^3{}_2 A^2{}_3 \end{aligned}$$

Theorem 6.1:

$$\sum_{s=1}^{N} A^s{}_q \operatorname{cof} A^s{}_t = \delta^t{}_q \det A \quad \text{and} \quad \sum_{t=1}^{N} A^q{}_t \operatorname{cof} A^s{}_t = \delta^q_s \det A \tag{6.35}$$

Proof: It follows from (6.34) that

$$\begin{aligned} \sum_{s=1}^{N} A^s{}_q \operatorname{cof} A^s{}_t &= \varepsilon_{i_1 i_2 \dots i_N} A^{i_1}{}_1 \dots A^{i_{t-1}}{}_{t-1} A^s{}_q \delta^{i_t}{}_s A^{i_{t+1}}{}_{t+1} \dots A^{i_N}{}_N \\ &= \varepsilon_{i_1 i_2 \dots i_N} A^{i_1}{}_1 \dots A^{i_{t-1}}{}_{t-1} A^{i_t}{}_q A^{i_{t+1}}{}_{t+1} \dots A^{i_N}{}_N \\ &= \varepsilon_{i_1 i_2 \dots i_N} A^{i_1}{}_1 \dots A^{i_{t-1}}{}_{t-1} A^{i_t}{}_t A^{i_{t+1}}{}_{t+1} \dots A^{i_N}{}_N \delta^t{}_q \\ &= \delta^t{}_q \det A \end{aligned}$$

Equation (6.35) follows by a similar argument.

Equations (6.35) represent the classical Laplace expansion of a determinant. Equations (6.35) fulfil this promise.

Exercises

1. Verify equation (6.20).
2. Verify equations (6.22)-(6.24).
3. Verify equations (6.25)-(6.27).

4. Show that the cofactor of an element of an $N \times N$ matrix A can be written

$$\text{cof } A^{i_1}{}_{j_1} = \frac{1}{(N-1)!}\delta^{j_1 \cdots j_N}_{i_1 \ldots i_N} A^{i_2}{}_{j_2} \ldots A^{i_N}{}_{j_N} \tag{6.36}$$

5. If A and B are $N \times N$ matrices use (6.24) to prove that

$$\det AB = \det A \det B$$

6. If I is the $N \times N$ identity matrix show that $\det I = 1$.
7. If A is a non-singular matrix show, that $\det A \neq 0$ and that

$$\det A^{-1} = \frac{1}{\det A}$$

8. If A is an $N \times N$ matrix, we define its $K \times K$ minor $(1 \le K \le N)$ to be the determinant of any $K \times K$ submatrix of A, e.g.,

$$A^{i_1 \ldots i_K}_{j_1 \cdots j_K} \equiv \det \begin{vmatrix} A^{i_1}{}_{j_1} & \cdot & \cdot & \cdot & A^{i_1}{}_{j_K} \\ \cdot & & & & \\ \cdot & & & & \\ \cdot & & & & \\ A^{i_K}{}_{j_1} & \cdot & \cdot & \cdot & A^{i_K}{}_{j_K} \end{vmatrix} = \delta^{k_1 \ldots k_K}_{j_1 \cdots j_K} A^{i_1}{}_{k_1} \ldots A^{i_K}{}_{k_K}$$

$$= \delta^{i_1 \ldots i_K}_{k_1 \ldots k_K} A^{k_1}{}_{j_1} \ldots A^{k_K}{}_{j_K} \tag{6.37}$$

In particular, any element $A^i{}_j$ of A is a 1×1 minor of A. The cofactor of the $K \times K$ minor $A^{i_1 \ldots i_K}_{j_1 \cdots j_K}$ is defined by

$$\text{cof } A^{i_1 \ldots i_K}_{j_1 \cdots j_K} \equiv \frac{1}{(N-K)!}\delta^{j_1 \cdots j_N}_{i_1 \cdots i_N} A^{i_{K+1}}{}_{j_{K+1}} \ldots A^{i_N}{}_{j_N} \tag{6.38}$$

which is equal to the $(N - K) \times (N - K)$ minor complementary to the $K \times K$ minor $A^{i_1 \ldots i_K}_{j_1 \cdots j_K}$ and is assigned an appropriate sign. Specifically, if $(i_{K+1}, \ldots, i_N)$ and $(j_{K+1}, \ldots, j_N)$ are complementary sets of $(i_1, \ldots, i_K)$ and $(j_1, \ldots, j_K)$ in $(1, \ldots, N)$, then

$$\operatorname{cof} A^{i_1 \ldots i_K}_{j_1 \ldots j_K} = \delta^{i_1 \ldots i_N}_{j_1 \ldots j_N} A^{i_{K+1} \ldots i_N}_{j_{K+1} \ldots j_N} \quad \text{(no summation)} \tag{6.39}$$

Clearly (6.38) generalises (6.36). Show that the formulas that generalise (6.35) to $K \times K$ minors in general are

$$\delta^{i_1 \ldots i_K}_{j_1 \ldots j_K} \det A = \frac{1}{K!} \sum_{k_1 \ldots k_K = 1}^{N} A^{i_1 \ldots i_K}_{k_1 \ldots k_K} \operatorname{cof} A^{j_1 \ldots j_K}_{k_1 \ldots k_K}$$

$$\delta^{j_1 \ldots j_K}_{i_1 \ldots i_K} \det A = \frac{1}{K!} \sum_{k_1 \ldots k_K = 1}^{N} A^{k_1 \ldots k_K}_{i_1 \ldots i_K} \operatorname{cof} A^{k_1 \ldots k_K}_{j_1 \ldots j_K} \tag{6.40}$$

9. If A is non-singular, show that

$$A^{i_1 \ldots i_K}_{j_1 \ldots j_K} = (\det A) \operatorname{cof} (A^{-1})^{j_1 \ldots j_K}_{i_1 \ldots i_K} \tag{6.41}$$

and that

$$\frac{1}{K!} \sum_{k_1 \ldots k_K = 1}^{N} A^{i_1 \ldots i_K}_{k_1 \ldots k_K} (A^{-1})^{k_1 \ldots k_K}_{j_1 \ldots j_K} = \delta^{i_1 \ldots i_K}_{j_1 \ldots j_K} \tag{6.42}$$

In particular, if $K = 1$ and A is replaced by A^{-1} then (6.41) reduces to

$$(A^{-1})^i_j = (\det A)^{-1} \operatorname{cof} A^j_i \tag{6.43}$$

Matrix of a Linear Transformation

In this section we shall introduce the matrix of a linear transformation with respect to a basis and investigate certain of its properties. Here we show how a certain matrix can be associated with a linear transformation, and more importantly, we show to what extent the relationship is basis dependent.

If $A \in L(V; U)$, $\{e_1, \ldots e_N\}$ is a basis for V, and $\{b_1, \ldots b_M\}$ is a basis for U, then we can characterise A by the system (5.63):

$$Ae_k = A^a_K b_a \tag{6.44}$$

where the summation is in force with the Greek indices ranging from 1 to M. The *matrix* of A with respect to the bases $\{e_1, \ldots e_N\}$ and $\{b_1, \ldots b_M\}$, denoted by $M(A, e_K, b_a)$ is

$$M(A, e_k, b_a) = \begin{vmatrix} A^1{}_1 & A^1{}_2 & \cdot & \cdot & \cdot & A^1{}_N \\ A^2{}_1 & A^2{}_2 & \cdot & \cdot & \cdot & A^2{}_N \\ \cdot & & & & & \\ \cdot & & & & & \\ \cdot & & & & & \\ A^M{}_1 & A^M{}_2 & \cdot & \cdot & \cdot & A^M{}_N \end{vmatrix} \tag{6.45}$$

As the above argument indicates, the matrix of A depends upon the choice of basis for V and U. However, unless this point needs to be stressed, we shall often write $M(A)$ for the matrix of A and the basis dependence is understood. We can always regard M as a function M: $L\ (U;\ V) \rightarrow M^{M\times N}$. It is a simple exercise to confirm that

$$M(\lambda A + \mu B) = \lambda M(A) + \mu M(B) \tag{6.46}$$

for all $\lambda, \mu \in C$ and $A, B \in L(U; V)$. Thus M is a linear transformation. Since

$$M(A)=0$$

implies $A = 0$, M is one-to-one. Since $L(U; V)$ and $M^{M\times N}$ have the same dimension, Theorem 5.10 tells us that M is an automorphism and, thus, $L\ (U; V)$ and $M^{M\times N}$ are isomorphic. The dependence of $M\ (A,\ e_{K'}\ b_a)$ on the bases can be exhibited by use of the transformation rule (5.67). In matrix form (5.67) is

$$M\ (A,\ e_{K'}\ b_a) = SM\ (A,\ \hat{e}_{K'}\ \hat{b}_a)T^{-1} \tag{6.47}$$

where S is the $M\times M$ matrix

$$S = \begin{vmatrix} S_1^1 & S_2^1 & \cdot & \cdot & \cdot & S_M^1 \\ S_1^2 & S_2^2 & \cdot & \cdot & \cdot & S_M^2 \\ \cdot & & & & & \\ \cdot & & & & & \\ \cdot & & & & & \\ S_1^M & S_2^M & \cdot & \cdot & \cdot & S_M^M \end{vmatrix} \tag{6.48}$$

and T is the $N\times M$ matrix

$$T = \begin{vmatrix} T_1^1 & T_2^1 & \cdot & \cdot & \cdot & T_N^1 \\ T_1^2 & T_2^2 & \cdot & \cdot & \cdot & T_N^2 \\ \cdot \\ \cdot \\ \cdot \\ T_1^N & T_2^N & \cdot & \cdot & \cdot & T_N^N \end{vmatrix} \tag{6.49}$$

Of course, in constructing (6.47) from (5.67) we have used the fact expressed by (4.44) and (4.45) that the matrix T^{-1} has components $\hat{T}_k^j, j, k = 1, \dots N$.

If A is an endomorphism, the transformation formula (6.47) becomes

$$\text{M}\ (A, e_k, e_q) = TM\ (A, \hat{e}_k, \hat{e}_q)\text{T}^{-1} \tag{6.50}$$

We shall use (6.50) to motivate the concept of the *determinant* of an endomorphism. The determinant of $M\ (A, e_k, e_q)$, written det $M\ (A, e_k, e_q)$, can be computed by (6.19). It follows from (6.50).

$$\det M\ (A, e_k, e_q) = (\det T)\ [(\det M\ (A, \hat{e}_k, \hat{e}_q)]\ (\det T^{-1})$$

$$= \det M\ (A, \hat{e}_k, \hat{e}_q)$$

Thus, we obtain the important result that det $M\ (A, e_k, e_q)$ is *independent* of the choice of basis for V. With this fact we define the determinant of an endomorphism $A \in L\ (V; V)$, written det A, by

$$\det A = \det \text{M}\ (A, e_k, e_q) \tag{6.51}$$

By the above argument, we are assured that det A is a property of A alone.

Given a linear transformation $A \in L\ (V, U)$, the adjoint $A^* \in L\ (U; V)^*$ is defined by the component formula (5.71). Consistent with equations (6.44) and (6.45), equation (5.71) implies that

$$M(A^*, b^\alpha, e^k) = \begin{vmatrix} {A^*}_1{}^1 & {A^*}_1{}^2 & \cdot & \cdot & \cdot & {A^*}_1{}^M \\ {A^*}_2{}^1 & {A^*}_2{}^2 & \cdot & \cdot & \cdot & {A^*}_2{}^M \\ \cdot \\ \cdot \\ \cdot \\ {A^*}_N{}^1 & {A^*}_N{}^2 & \cdot & \cdot & \cdot & {A^*}_N{}^M \end{vmatrix} \tag{6.52}$$

Notice that the matrix of A^* is referred to the reciprocal bases $\{e^k\}$ and $\{b^\alpha\}$. If we now use (5.72) and the definition (6.45) we see that

$$M(A^*, b^\alpha, e^k) = \overline{M(A, e_k, b_\alpha)^T} \tag{6.53}$$

where the complex conjugate of a matrix is the matrix formed by taking the complex conjugate of each component of the given matrix. Equation (6.53) gives a simple comparison of the component matrices of a linear transformation and its adjoint. If the vector spaces are real, (6.53) reduces to

$$M\ (A^T) = M\ (A)^T \tag{6.54}$$

where the basis dependence is understood. For an endomorphism $A \in L\ (V;\ V)$, we can use (6.53) and (6.51) to show that

$$\det A^* = \overline{\det A} \tag{6.55}$$

Given a $M \times N$ matrix $A = [A^\alpha{}_k]$, there correspond M $1 \times N$ row matrices and N $M \times 1$ column matrices. The *row rank* of the matrix A is equal to the number of linearly independent row matrices and the *column rank* of A is equal to the number of linearly independent column matrices. It is a property of matrices, which we shall prove, that the row rank *equals* the column rank. This common rank in turn is equal to the rank of the linear transformation whose matrix is A. The theorem we shall prove is the following.

Theorem 6.2: $A = [A^\alpha{}_k]$, is an $M \times N$ matrix, the row rank of A equals the column rank of A.

Proof: Let $A \in L\ (V;\ U)$ and let $M\ (A) = A = [A^\alpha{}_k]$ with respect to bases $\{e_1,\ldots,e_N\}$ for V and $\{b_1,\ldots,b_M\}$ for U. We can define a linear transformation $B : U \to C^M$ by

$$Bu = (u^1, u^2, \ldots, u^M)$$

where $u = u^\alpha b_\alpha$. Observe that B is an isomorphism. The product BA is a linear transformation $V \to C^M$; further,

$$BAe_k = B(A^\alpha{}_k b_\alpha) = A^\alpha{}_k B b_a = (A^1{}_k, A^2{}_k, \ldots, A^M{}_k)$$

Therefore, BAe_k is an M-tuple whose elements are those of the kth column matrix of A. This means dim $R(BA)$ = column rank of A. Since B is an isomorphism, BA and A have the same rank and thus column rank of A = dim $R(A)$. A similar argument applied to the adjoint mapping A^* shows that row rank of A = dim $R(A^*)$. If we now apply Theorem 5.20 we find the desired result.

In what follows we shall only refer to the rank of a matrix rather than to the row or the column rank.

Theorem 6.3: An endomorphism A is regular if and only if det $A \neq 0$.

Proof: ~ If det $A \neq 0$, equation (6.25) provide us with a formula for the direct calculation of A^{-1} so A is regular. If A is regular, then A^{-1} exists, and show that det $A \neq 0$.

Before closing this section, we mention here that among the four component matrices $[A^{\alpha}{}_{k}], [A_{\alpha k}], [A^{\alpha k}]$, and $[A_{\alpha}{}^{k}]$ defined by (5.63), (5.78), (5.79), and (5.80), respectively, the first one, $[A_{\alpha}{}^{k}]$, is most frequently used. For example, it is that particular component matrix of an endomorphism A that we used to define the determinant of A, as shown in (6.51). In the sequel, if we refer to the component matrix of a transformation or an endomorphism, then unless otherwise specified, we mean the component matrix of the first kind.

Exercises

1. If A and B are endomorphisms, show that

   $M(AB) = M(A)\,M(B)$ and $\det AB = \det A \det B$

2. If A: $V \rightarrow V$ is a skew-Hermitian endomorphism, show that

   $$\det A = (-1)^{\dim V}\,\overline{\det A}$$

3. If $A \in L(V;V)$ is regular, show that

   $M(A^{-1}) = M(A)^{-1}$

4. If $P \in L(V;V)$ is a projection, show that there exists a basis for V such that

$$M(P)=\left|\begin{array}{cccccccc} 1 & & & & & & & \\ & 1 & & & & & & \\ & & \ddots & & 0 & & & \\ & & & 1 & & & & \\ & & & & 0 & & & \\ & & & & & 0 & & \\ & 0 & & & & & \ddots & \\ & & & & & & & 0 \end{array}\right| \begin{array}{l} \left.\begin{array}{c} \\ \\ \\ \end{array}\right\} R=\dim R(P) \\ \left.\begin{array}{c} \\ \\ \\ \end{array}\right\} N-R \end{array}$$

5. If $A \in L(V;V)$ is Hermitian, show that the component version of this fact is

$$A_{ij}=\overline{A_{ji}}, \qquad A^i{}_j=\overline{A^i_j}, \qquad \text{and} \qquad e_{ik}A^i{}_j=\overline{e_{ji}A_k{}^i}$$

where $e_{ik} = e_i \cdot e_k$, and where $\{e_1,\dots,e_N\}$ is the basis used to express the transformation formulas (5.78)-(5.80) in matrix notation.

6. If $A \in L(V;V)$ we have seen it has four sets of components which are denoted by A_i^j, $A^i{}_j$, A^{ij}, and A_{ij}. Prove that det $[A_{ij}]$ and det $[A^{ij}]$ *do* depend upon the choice of basis for V.

7. Show that

$$\det(\alpha A)=\alpha^N \det A \tag{6.56}$$

for all scalars α. In particular, if $A = I$, then

$$\det(\alpha I)=\alpha^N \tag{6.57}$$

8. If A is an endomorphism of an N-dimensional vector space V, show that det $(A-\lambda I)$ is a polynomial of degree N in λ.

Systems of Linear Equations

In this section we shall examine the problem system of M equations in N unknowns of the form

$$A^\alpha{}_k v^k = u^\alpha, \qquad \alpha=1,\dots,M, \quad k=1,\dots,N \tag{6.58}$$

where the MN coefficients $A^{\alpha}{}_{k}$ and the M data u^{α} are given. If we introduce bases $\{e_1,...,e_N\}$ for a vector space V and $\{b_1,...,b_M\}$ for a vector space U, then (6.58) can be viewed as the component formula of a certain vector equation

$$Av = u \tag{6.59}$$

which immediately yields the following theorem.

Theorem 6.4: (6.58) has a solution if and only if u is in $R(A)$.

Another immediate implication of (6.59) is the following theorem.

Theorem 6.5: If (6.58) has a solution, the solution is unique if and only if A is regular.

Given the system of equations (6.58), the *associated homogeneous system* is the set of equations

$$A^{\alpha}{}_{k}v^{k} = 0 \tag{6.60}$$

Theorem 6.6: The set of solutions of the homogeneous system (6.60) whose coefficient matrix is of rank R form a vector space of dimension N - R.

Proof: Equation (6.60) can be regarded as the component formula of the vector equation

$$Av = 0$$

which implies that v^k solves (6.60) if and only if $v \in K(A)$. By (5.6) dim $K(A)$ = div V - dim $R(A)$ = N - R.

From Theorems 5.7 and 6.6 we see that if there are fewer equations than unknowns (i.e. $M < N$), the system (6.60) always has a non-zero solution. This assertion is clear because $N - R \geq N - M > 0$, since $R(A)$ is a subspace of U and M = dim U.

If in (6.58) and (6.60) $M = N$, the system (6.58) has a solution for all u^k if and only if (6.60) has the trivial solution $v^1 = v^2 = ... = v^N = 0$ only. For, in this circumstance, A is regular 0 and thus invertible. This means det $[A^{\alpha}{}_{k}] \neq 0$, and we can use (6.25) and write the solution of (6.58) in the form

$$v^j = \frac{1}{\det[A^\alpha{}_k]} \sum_{\alpha=1}^{N} (\mathrm{cof}\, A^\alpha{}_j) u^\alpha \tag{6.61}$$

which is the classical Cramer's rule

Exercise

1. Given the system of equations (6.58), the *augmented matrix* of the system is the matrix obtained from $[A^\alpha{}_k]$ by the addition of a column formed from $u^1, \ldots u^N$. Use Theorem 6.4 and prove that the system (6.58) has a solution if and only if the rank of $[A^\alpha{}_k]$ equals the rank of the augmented matrix.

Spectral Decompositions

In this chapter we consider one of the more advanced topics in the study of linear transformations. Essentially we shall consider the problem of analysing an endomorphism by decomposing it into elementary parts.

Direct Sum of Endomorphisms

If A is an endomorphism of a vector space V, a subspace V_1 of V is said to be A-invariant if A maps V_1 to V_1. The most obvious example of an A-invariant subspace is the null space $K(A)$. Let $A_1, A_2, \ldots, A_L$ be endomorphisms of V; then an endomorphism A is the *direct sum* of $A_1, A_2, \ldots, A_L$ if,

$$A = A_1 + A_2 + \cdots + A_L \tag{7.1}$$

and

$$A_i A_j = 0 \qquad i \neq j \tag{7.2}$$

For example, equation (5.35) presents a direct sum decomposition for the identity linear transformation.

Theorem 7.1: If $A \in L\,(V;U)$ and $V = V_1 \oplus V_2 \oplus \cdots \oplus V_L$, where each subspace V_j is A-invariant, then A has the direct sum decomposition (7.1), where each A_t is given by

$$A_i v_i = A v_i \tag{7.3(a)}$$

for all $v_i \in V_i$ and

$$A_i v_j = 0 \tag{7.3(b)}$$

For all $v_j \in V_j$, $i \neq j$, for $i = 1,\ldots,L$. Thus, the restriction of A to V_j coincides with that of A_j; further, each V_i is A_j-invariant, for all $j = 1,\ldots,L$.

Proof: Given the decomposition $V = V_1 \oplus V_2 \oplus \cdots \oplus V_L$, then $v \in V$ has the unique representation $v = v_1 + \cdots + v_L$, where each $v_i \in V_i$. By the converse of Theorem 5.16, there exist L projections $P_1,\ldots,P_L$ (5.70) and also $v_j \in P_j(V)$. We have $A_j = AP_j$, and since V_j is A - invariant, $AP_j = P_j A$. Therefore, if $i \neq j$,

$$A_i A_j = AP_i AP_j = AAP_i P_j = 0$$

where (5.34) has been used. Also

$$A_1 + A_2 + \cdots + A_L = AP_1 + AP_2 + \cdots + AP_L$$
$$= A(P_1 + P_2 + \cdots + P_L)$$
$$= A$$

where (5.35) has been used.

When the assumptions of the preceding theorem are satisfied, the endomorphism A is said to be *reduced* by the subspaces $V_1,\ldots, V_L$. An important result of this circumstance is contained in the following theorem.

***Theorem** 7.2:* Under the conditions of Theorem 7.1, the determinant of the endomorphism A is given by

$$\det A = \det A_1 \det A_2 \cdots \det A_L \tag{7.4}$$

where A_k denotes the restriction of A to V_k for all $k = 1,\ldots,L$.

The proof of this theorem is left as an exercise to the reader.

Exercises

1. Assume that $A \in L(V; V)$ is reduced by $V_1,\ldots, V_L$ and select for the basis of V the union of some basis for the subspaces

$V_1,\ldots,V_L$. Show that with respect to this basis M (A) has the following *block* form:

$$M(A) = \begin{bmatrix} A_1 & & & 0 \\ & A_2 & & \\ & & \ddots & \\ 0 & & & A_L \end{bmatrix} \tag{7.5}$$

where A_j is the matrix of the restriction of A to V_j.

2. Use the result of Exercise .1 and prove Theorem 7.2.

3. Show that if $V = V_1 \oplus V_2 \oplus \cdots \oplus V_L$, then $A \in L(V;V)$ is *reduced* by $V_1 \ldots V_L$ if and only if $AP_j = P_jA$ for $j = 1,\ldots,L$, where P_j is the projection on V_j given by (5.38).

Eigenvectors and Eigenvalues

Given an endomorphism $A \in L(V;V)$, the problem of finding a direct sum decomposition of A is closely related to the study of the spectral properties of A. This concept is central in the discussion of *eigenvalue problems*.

A scalar λ is an eigenvalue of $A \in L(V;V)$ if there exists a non-zero vector $v \in V$ such that

$$Av = \lambda v \tag{7.6}$$

The vector v in (7.6) is called an *eigenvector* of A corresponding to the eigenvalue λ. Eigenvalues and eigenvectors are sometimes refered to as *latent roots* and *latent vectors, characteristic roots* and *characteristic vectors*, and *proper values* and *proper vectors*. The set of all eigenvalues of A is the spectrum of A, denoted by $\sigma(A)$. For any $\lambda \in \sigma(A)$, the set

$$V(\lambda) = \{v \in V \mid Av = \lambda v\}$$

is a subspace of V, called the *eigenspace* or *characteristic subspace* corresponding to λ. The geometric multiplicity of λ is the dimension of $V(\lambda)$.

Theorem 7.3: Given any $\lambda \in \sigma(A)$, the corresponding eigenspace $V(\lambda)$ is A-invariant.

The proof of this theorem involves an elementary use of (7.6). Given any $\lambda \in \sigma(A)$, the restriction of v to $V(\lambda)$, denoted by $A_{V(\lambda)}$, has the property that

$$A_{V(\lambda)} u = \lambda u \tag{7.7}$$

for all u in $V(\lambda)$. Geometrically, $A_{V(\lambda)}$ simply amplifies u by the factor λ. Such lineare transformations are often called *dilatations.*

Theorem 7.4: $A \in L(V;V)$ is not regular if and only if 0 is an eigenvalue of A; further, the corresponding eigenspace is $K(A)$.

The proof is obvious. Note that (7.6) can be written as $(A - \lambda I)$ $v = 0$, which shows that

$$V(\lambda) = K(A - \lambda I) \tag{7.8}$$

Equation (7.8) implies that λ is an eigenvalue of A if and only if $A - \lambda I$ is singular. Therefore, by Theorem 6.3,

$$\lambda \in \sigma(A) \Leftrightarrow \det(A - \lambda I) = 0 \tag{7.9}$$

The polynomial $f(\lambda)$ of degree N = dim V defined by

$$f(\lambda) = \det(A - \lambda I) \tag{7.10}$$

is called the *characteristic polynomial* of A. Equation (7.10) shows that the eigenvalues of A are roots of the characteristic equation

$$f(\lambda) = 0 \tag{7.11}$$

Exercises

1. Let A be an endomorphism whose matrix $M(A)$ relative to a particular basis is a triangular matrix, say

$$M(A) = \begin{bmatrix} A^1{}_1 & A^1{}_2 & \cdot & \cdot & \cdot & A^1{}_N \\ & A^2{}_2 & & & & \\ & & \cdot & & & \cdot \\ & & & \cdot & & \cdot \\ & & & & \cdot & \cdot \\ 0 & & & & & A^N{}_N \end{bmatrix}$$

What is the characteristic polynomial of A? Use (7.11) and determine the eigenvalues of A.

2. Show that the eigenvalues of a Hermitian endomorphism of an inner product space are all real numbers and the eigenvalues of a skew-Hermitian endomorphism are all purely imaginary numbers (including the number $0 = i0$).
3. Show that the eigenvalues of a unitary endomorphism all have absolute value 1 and the eigenvalues of a projection are either 1 or 0.
4. Show that the eigenspaces corresponding to different eigenvalues of a Hermitian endomorphism are mutually orthogonal.
5. If $B \in L(V;V)$ is regular, show that $B^{-1}AB$ and A have the same characteristic equation. How are the eigenvectors of $B^{-1}AB$ related to those of A?

Characteristic Polynomial

The eigenvalues of $A \in L$ (V;V) are roots of the characteristic polynomial

$$f(\lambda) = 0 \tag{7.12}$$

where

$$f(\lambda) = \det(A - \lambda I) \tag{7.13}$$

If $\{e_1, \ldots, e_2\}$ is a basis for V, then by (5.68)

$$Ae_k = A^j{}_k e_j \tag{7.14}$$

Therefore, by (7.13) and (6.27),

$$f(\lambda) = \frac{1}{N!}\delta^{j_1 \cdots j_N}_{i_1 \cdots i_N}\left(A^{i_1}{}_{j_1} - \lambda\delta^{i_1}_{j_1}\right)\cdots\left(A^{i_N}{}_{j_N} - \lambda\delta^{i_N}_{j_N}\right) \tag{7.15}$$

If (7.15) is expanded and we use (6.11), the result is

$$f(\lambda) = (-\lambda)^N + \mu_1(-\lambda)^{N-1} + \cdots + \mu_{N-1}(-\lambda) + \mu_N \tag{7.16}$$

where

$$\mu_j = \frac{1}{j!}\delta^{q_1,\ldots,q_j}_{i_1,\ldots,i_j} A^{i_1}{}_{q_1} \cdots A^{i_j}{}_{q_j} \tag{7.17}$$

Since $f(\lambda)$ is defined by (7.13), the coefficients μ_j, $j = 1,\ldots,N$, are independent of the choice of basis for V. These coefficients are called the *fundamental invariants* of A. Equation (7.17) specialises to

$$\mu_1 = tr\ A,\ \mu_2 = \frac{1}{2}\left\{(tr\ A)^2 - tr\ A^2\right\},\ \text{and}\ \mu_N = det\ A \tag{7.18}$$

where (6.20) has been used to obtain (7.18). Since $f(\lambda)$ is a Nth degree polynomial, it can be factored into the form

$$f(\lambda) = (\lambda_1 - \lambda)^{d_1}(\lambda_2 - \lambda)^{d_2} \cdots (\lambda_L - \lambda)^{d_L} \tag{7.19}$$

where $\lambda_1,\ldots,\lambda_L$ are the distinct foots of $f(\lambda) = 0$ and $d_1,\ldots, d_L$ are positive integers which must satisfy $\sum_{j=1}^{L} d_j = N$. It is in writing (7.19) that we have made use of the assumption that the scalar field is complex. If the scalar field is real, the polynomial (7.16), generally, cannot be factored.

In general, a scalar field is said to be algebraically closed if every polynomial equation has at least one root in the field, or equivalently, if every polynomial, such as $f(\lambda)$, can be factored into the form (7.19). It is a theorem in algebra that the complex field is algebraically closed. The real field, however, is not algebraically closed. For example, if λ is real the polynomial equation $f(\lambda) = \lambda^2 + 1 = 0$ has no real roots. By allowing the scalar fields to be complex, we are assured that every endomorphism has at least one eigenvector. In the expression (7.19), the integer d_j is called the *algebraic multiplicity* of the eigenvalue λ_j. It is possible to prove that the algebraic multiplicity of an eigenvalue is not less than the geometric multiplicity of the same eigenvalue.

An expression for the invariant μ_j can be obtained in terms of the eigenvalues if (7.19) is expanded and the results are compared to (7.16). For example,

$$\mu_1 = tr\ A = d_1\lambda_1 + d_2\lambda_2 + \cdots + d_L\lambda_L \tag{7.20}$$

$$\mu_N = \det A = \lambda_1^{d_1}\lambda_2^{d_2}\cdots\lambda_L^{d_L}$$

The next theorem we want to prove is known as the *Cayley-Hamilton* theorem. Roughly speaking, this theorem asserts that an endomorphism satisfies its own characteristic equation. To make this statement clear, we need to introduce certain ideas associated with polynomials of endomorphisms. If $A \in L(V;V)$, A^2 is defined by

$$A^2 = AA$$

Similarly, we define by induction, starting from

$$A^0 = I$$

and in general

$$A^k = AA^{k-1} = A^{k-1}A$$

where k is any integer greater than one. If k and l are positive integers, it is easily established by induction that

$$A^k A^l = A^l A^k = A^{l+k} \tag{7.21}$$

Thus, A^k and A^l commute. *A polynomial* in A is an endomorphism of the form

$$g(A) = \alpha_0 A^M + \alpha_1 A^{M-1} + \cdots + \alpha_{M-1}A + \alpha_M I \tag{7.22}$$

where M is a positive integer and $\alpha_0, \ldots, \alpha_M$ are scalars. Such polynomials have certain of the properties of polynomials of scalars. For example, if a scalar polynomial

$$g(t) = \alpha_0 t^M + \alpha_1 t^{M-1} + \cdots + \alpha_{M-1}t + \alpha_M$$

can be factored into

$$g(t) = \alpha_0 (t - \eta_1)(t - \eta_2)\cdots(t - \eta_M)$$

then the polynomial (7.22) can be factored into

$$g(A) = \alpha_0 (A - \eta_1 I)(A - \eta_2 I)\cdots(A - \eta_M I) \tag{7.23}$$

The order of the factors in (7.23) is not important since, as a result of (7.21), the factors commute. Notice, however, that the product of two non-zero endomorphisms can be zero. Thus, the formula

$$g_1(A)g_2(A) = 0 \tag{7.24}$$

for two polynomials g_1 and g_2 generally *does not* imply one of the factors is zero. For example, any projection P satisfies the equation $P(P-I) = 0$, but generally P and $P-I$ are both non-zero.

Theorem 7.5: (Cayley-Hamilton). If $f(\lambda)$ is the characteristic polynomial (7.16) for an endomorphism A , then

$$f(A) = (-A)^N + \mu_1(-A)^{N-1} + \cdots + \mu_{N-1}(-A) + \mu_N I = 0 \tag{7.25}$$

Proof: The proof which we shall now present makes major use of equation (1.29) or, equivalently, (6.35). If $\operatorname{adj}(A-\lambda I)$ is the endomorphism whose matrix is $\operatorname{adj}\left[A^p{}_q - \lambda\delta^p_q\right]$, where $\left[A^p{}_q\right] = M$ (A), then by (7.13) and (1.29)

$$(\operatorname{adj}(A-\lambda I))\,(A-\lambda I) = f(\lambda)I \tag{7.26}$$

By (6.34) it follows that $\operatorname{adj}(A-\lambda I)$ is a polynomial of degree $N - 1$ in λ. Therefore,

$$adj(A-\lambda I) = B_0(-\lambda)^{N-1} + B_1(-\lambda)^{N-2} + \cdots + B_{N-2}(-\lambda) + B_{N-1} \tag{7.27}$$

where $B_0,\ldots,B_{N-1}$ are endomorphisms determined by A. If we now substitute (7.27) and (7.26) into (7.25) and require the result to hold for all λ, we find

$$\begin{aligned}
B_0 &= I \\
B_0 A + B_1 &= \mu_1 I \\
B_1 A + B_2 &= \mu_2 I \\
&\;\;\vdots \\
B_{N-2}A + B_{N-1} &= \mu_{N-1} I \\
B_{N-1}A &= \mu_N I
\end{aligned} \tag{7.28}$$

Now we multiply $(7.28)_1$ by $(-A)^N$, $(7.28)_2$ by $(-A)^{N-1}$, $(7.28)_3$ by $(-A)^{N-2}$,...,$(7.28)_k$ by $(-A)^{N-k+1}$, etc., and add the resulting N equations, to find

$$(-A)^N + \mu_1(-A)^{N-1} + \cdots + \mu_{N-1}(-A) + \mu_N I = 0 \tag{7.29}$$

which is the desired result.

Exercises

1. If $N = \dim V$ is odd, and the scalar field is R, prove that the characteristic equation of any $A \in L(V;V)$ has at least one eigenvalue.
2. If $N = \dim V$ is odd and the scalar field is R, prove that if $\det A > 0$ (< 0), then A has at least one positive (negative) eigenvalue.
3. If $N = \dim V$ is even and the scalar field is R and $\det A < 0$, prove that A has at least one positive and one negative eigenvalue.
4. $A \in L(V;V)$ be regular. Show that

   $\det(A^{-1} - \lambda^{-1} I) = (-\lambda)^{-N} \det A^{-1} \det(A - \lambda I)$

   where $N = \dim V$.
5. Prove that the characteristic polynomial of a projection $P : V \to V$ is

   $f(\lambda) = (-1)^{-N} \lambda^{N-L}(1-\lambda)$

   where L = dim V and $L = \dim P(V)$.
6. If $A \in L(V;V)$ is regular, express A^{-1} as a polynomial in A.
7. Prove directly that the quantity μ_j defined by (7.17) is independent of the choice of basis for V.
8. Let C be an endomorphism whose matrix relative to a basis $\{e_1,...,e_N\}$ is a triangular matrix of the form

$$M(C)=\begin{bmatrix} 0 & 1 & 0 & \cdot & \cdot & \cdot & 0 \\ & 0 & 1 & 1 & & & \\ & & \cdot & \cdot & \cdot & & \\ & & & \cdot & \cdot & \cdot & \\ & & & & \cdot & \cdot & 0 \\ & & & & & 0 & 1 \\ 0 & & & & & & 0 \end{bmatrix} \tag{7.30}$$

i.e. C maps e_1 to 0, e_2 to e_1, e_3 to e_2,.... Show that a necessary and sufficient condition for the existence of a component matrix of the form (7.30) for an endomorphism C is that

$$C^N = 0 \text{ but } C^{N-1} \neq 0 \tag{7.31}$$

We call such an endomorphism C a *nilcyclic* endomorphism and we call the basis $\{e_1,...,e_N\}$ for the form (7.30) a *cyclic basis* for C.

9. If $\phi_1,...,\phi_N$ denote the fundamental invariants of A^{-1}, use the results of Exercise 4 to show that

$$\phi_j = \left(\det A^{-1}\right)\mu_{N-j}$$

for $j = 1,..., N$.

10. It follows from (7.27) that

$$B_{N-1} = \text{adj } A$$

Use this result along with (7.28) and show that

$$\text{adj } A = (-A)^{N-1} + \mu_1 (-A)^{N-2} + ... + \mu_{N-1} I$$

11. Show that

$$\mu_{N-1} = \text{tr}(\text{adj}A)$$

and, from Exercise 10, that

$$\mu_{N-1} = -\frac{1}{N-1}\left\{tr(-A)^{N-1} + \mu_1 tr(-A)^{N-2} + \cdots + \mu_{N-2} tr(-A)\right\}$$

Spectral Decomposition

An important problem in mathematical physics is to find a basis for a vector space V in which the matrix of a given $A \in L(V;V)$ is diagonal. If we examine (7.6), we see that if V has a basis of eigenvectors of A, then $M(A)$ is diagonal and vice versa; further, in this case the diagonal elements of $M(A)$ are the eigenvalues of A.

As we shall see, not all endomorphisms have matrices which are diagonal. Rather than consider this problem in general, we specialise here to the case where A,. then $M(A)$ is Hermitian and show that every Hermitian endomorphism has a matrix which takes on the diagonal form. First, we prove a general theorem for arbitrary endomorphisms about the linear independence of eigenvectors.

Theorem 7.6: If $\lambda_1, ..., \lambda_l$ are distinct eigenvalues of $A \in L(V;V)$ and if $u_1,...,u_L$ are eigenvectors corresponding to them, then $\{u_1,...,u_L\}$ form a linearly independent set.

Proof: If $\{u_1,...,u_L\}$ is not linearly independent, we choose a maximal, linearly independent subset, say $\{u_1,...,u_S\}$, from the set $\{u_1,...,u_L\}$; then the remaining vectors can be expressed uniquely as linear combinations of $\{u_1,...,u_S\}$, say

$$u_{S+1} = \alpha_1 u_1 + \cdots + \alpha_S u_S \tag{7.32}$$

where $\alpha_1,...,\alpha_S$ are not all zero and unique, because $\{u_1,...,u_S\}$ is linearly independent. Applying A to (7.32) yields

$$\lambda_{S+1} u_{S+1} = (\alpha_1 \lambda_1) u_1 + \cdots + (\alpha_S \lambda_S) u_S \tag{7.33}$$

Now if $\lambda_{S+1} = 0$ then $\lambda_1, ..., \lambda_S$ are non-zero because the eigenvalues are distinct and (7.33) contradicts the linear independence of $\{u_1,...,u_S\}$; on the other hand, if $\lambda_{S+1} \neq 0$, then we can divide (7.33) by λ_{S+1}, obtaining another expression of u_{S+1} as a linear combination of $\{u_1,...,u_S\}$ contradicting the uniqueness of the coefficients $\alpha_1,...,\alpha_S$. Hence, in any case the maximal linearly independent subset cannot be a proper subset of $\{u_1,...,u_L\}$; thus $\{u_1,...,u_L\}$ is linearly independent.

As a corollary to the preceding theorem, we see that if the geometric multiplicity is equal to the algebraic multiplicity for each eigenvalue of A, then the vector space V admits the direct sum representation

$$V = V(\lambda_1) \oplus V(\lambda_2) \oplus \cdots \oplus V(\lambda_L)$$

where $\lambda_1, \ldots, \lambda_L$ are the distinct eigenvalues of A. The reason for this representation is obvious, since the right-hand side of the above equation is a subspace having the same dimension as V; thus that subspace is equal to V. Whenever the representation holds, we can always choose a basis of V formed by bases of the subspaces $V(\lambda_1), \ldots, V(\lambda_L)$. Then this basis consists entirely of eigenvectors of A becomes a diagonal matrix, namely,

$$M(A) = \begin{bmatrix} \lambda_1 & & & & & & & & \\ & \cdot & & & & & & & \\ & & \cdot & & & & & 0 & \\ & & & \lambda_1 & & & & & \\ & & & & \lambda_2 & & & & \\ & & & & & \cdot & & & \\ & & & & & & \cdot & & \\ & & & & & & & \lambda_2 & \\ & & & & & & & & \cdot \\ & & & & & & & & & \cdot \\ & & & & & & & & & & \lambda_L \\ & 0 & & & & & & & & & & \cdot \\ & & & & & & & & & & & & \cdot \\ & & & & & & & & & & & & & \lambda_L \end{bmatrix} \qquad [7.34(a)]$$

where each λ_K is repeated d_k times, d_k being the algebraic as well as the geometric multiplicity of λ_K. Of course, the representation of V by direct sum of eigenspaces of A is possible if A has $N = \dim V$ distinct eigenvalues. In this case, the matrix of A taken with respect to a basis of dim eigenvectors has the diagonal form

$$M(A) = \begin{bmatrix} \lambda_1 & & & & & & \\ & \lambda_2 & & & 0 & & \\ & & \lambda_3 & & & & \\ & & & \cdot & & & \\ & & & & \cdot & & \\ & 0 & & & & \cdot & \\ & & & & & & \lambda_N \end{bmatrix} \qquad [7.34(b)]$$

If the eigenvalues of v are not all distinct, then in general the geometric multiplicity of an eigenvalue may be less then the algebraic multiplicity. Whenever the two multiplicities are different for at least one eigenvalue of A, it is no longer possible to find any basis in which the matrix of\ is diagonal. However, if V is an inner product space and if A is Hermitian, then a diagonal matrix of A can always be found; we shall now investigate this problem.

Recall that if u and v are arbitrary vectors in V, the adjoint A^* of $A \in L\ (V;V)$ defined by

$$Au\ .\ v = u\ .\ A^*v \tag{7.35}$$

As usual, if the matrix of A is referred to a basis $\{ e_k \}$, then the matrix of A^* is referred to the reciprocal basis $\{e^k\}$ and is given by [cf. (5.43)]

$$M\left(A^*\right) = \overline{M(A)^T} \tag{7.36}$$

where the overbar indicates the complex conjugate as usual. If A Hermitian, i.e., if $A = A^*$, then (7.35) reduces to

$$Au\ .\ v = u\ .\ Av \tag{7.37}$$

for all $u,\ v \in V$.

***Theorem** 7.7:* The eigenvalues of a Hermitian endomorphism are all real.

Proof: Let $A \in L\ (V;V)$ be Hermitian. Since $Au = \lambda\, u$ for any eigenvalue λ, we have

$$\lambda = \frac{Au \cdot u}{u \cdot u} \tag{7.38}$$

Therefore, we must show thatAu $\cdot$ u is real or, equivalently, we must show $Au \cdot u = \overline{Au \cdot u}$. By (7.37)

$$Au \cdot u = u \cdot Au = \overline{Au \cdot u}$$

where the rule $u \cdot v = \overline{v \cdot u}$ has been used. Equation (7.39) yields the desired result.

Theorem 7.8: If A is Hermitian and if V_1 is an A-invariant subspace of V, then $V_1^{\perp}$ is also A-invariant.

Proof: If $v \in V_1$ and $u \in V_1^{\perp}$, then $Av \cdot u = 0$ because $Av \in V_1$. But since A is Hermitian, $Av \cdot u = v \cdot Au = 0$. Therefore, $A\,u \in V_1^{\perp}$, which proves the theorem.

Theorem 7.9: If A is Hermitian, the algebraic multiplicity of each eigenvalue equals the geometric multiplicity.

Proof: Let $V(\lambda_0)$ be the characteristic subspace associated with an eigenvalue λ_0. Then the geometric multiplicity of λ_0 is $M = \dim\ V(\lambda_0)$. By Theorems 4.32 and 7.34

$$V - V(\lambda_0) \oplus V(\lambda_0)^{\perp} \tag{7.40}$$

Where $V(\lambda_0)$ and $V(\lambda_0)^{\perp}$ are A-invariant. By Theorem 7.1

$$A = A_1 + A_2$$

and by Theorem 5.16

$$I = P_1 + P_2$$

where P_1 projects V onto $V(\lambda_0)$, P_2 projects V onto $V(\lambda_0)^{\perp}$, $A_1 = AP_1$, and $A_2 = AP_2$. By Theorem 5.26, P_1 and P_2 and are Hermitian and they also commute with A. Indeed, for any $v \in V$, $P_1 v \in V(\lambda_0)$ and $P_1 v \in V(\lambda_0)^{\perp}$, and thus $AP_1 v \in V(\lambda_0)$ and $AP_2 \in V(\lambda_0)^{\perp}$. But since

$$Av = A(P_1+P_2)v = AP_1 v + AP_2 v$$

we see that $AP_1 v$ is the $V(\lambda_0)$ component of Av and $AP_2 v$ is the $V(\lambda_0)^{\perp}$ component of Av. Therefore

$$P_1 Av = AP_1 v, \qquad P_2 Av = AP_2 v$$

for all $v \in V$, or, equivalently

$$P_1 A = AP_1, \qquad P_2 A = AP_2$$

Together with the fact that P_1, and P_2 are Hermitian, these equations imply that A_2 and A_2 are also Hermitian. Further, A is reduced by the subspaces $V(\lambda_0)$ and $V(\lambda_0)^{\perp}$, since

$$A_1 A_2 = AP_1 AP_2 = A^2 P_1 P_2$$

Thus, if we select a basis $\{e_1,\ldots,e_N\}$ such that $\{e_1,\ldots,e_M\}$ span $V(\lambda_0)$ and $\{e_{M+1},\ldots,e_N\}$ span $V(\lambda_0)^{\perp}$, then the matrix of A to $\{e_k\}$ takes the form

$$M(A)=\left[\begin{array}{ccc|ccc} A^1{}_1 & \cdots & A^1{}_M & & & \\ \vdots & & \vdots & & 0 & \\ A^M{}_1 & \cdots & A^M{}_M & & & \\ \hline & & & A^{M+1}{}_{M+1} & \cdots & A^{M+1}{}_N \\ & 0 & & \vdots & & \vdots \\ & & & A^N{}_{M+1} & \cdots & A^N{}_N \end{array}\right]$$

and the matrices of A_1 and A_2 are

$$M(A_1)=\left[\begin{array}{ccc|c} A^1{}_1 & \cdots & A^1{}_M & \\ \vdots & & \vdots & 0 \\ A^M{}_1 & \cdots & A^M{}_M & \\ \hline & 0 & & 0 \end{array}\right]$$

$$M(A_2)=\left[\begin{array}{c|ccc} 0 & & 0 & \\ \hline & A^{M+1}{}_{M+1} & \cdots & A^{M+1}{}_N \\ 0 & \vdots & & \vdots \\ & A^N{}_{M+1} & \cdots & A^N{}_N \end{array}\right] \tag{7.41}$$

which imply

$$\det(A-\lambda I)=\det(A_1-\lambda I_{V(\lambda_0)})\det(A_2-\lambda I_{V(\lambda_0)^{\perp}})$$

By (7.7), $A_1 = \lambda_0 I_{V(\lambda_0)}$; thus by (6.37)

$$\det(A - \lambda I) = \det(\lambda_0 - \lambda)^M \det(A_2 - \lambda I_{V(\lambda_0)^\perp})$$

On the other hand, λ_0 is not an eigenvalue of A_2. Therefore

$$\det(A_2 - \lambda I_{V(\lambda_0)^\perp}) \neq 0$$

Hence, the algebraic multiplicity of λ_0 equals M, the geometric multiplicity.

The preceding theorem implies immediately the important result that V can be represented by a direct sum of eigenspaces of A if A is Hermitian. In particular, there exists a basis consisting entirely in eigenvectors of A, and the matrix of A relative to this basis is diagonal. However, before we state this result formally as a theorem, we first strengthen the result of Theorem 7.6 for the special case that A is Hermitian.

Theorem 7.10: If A is Hermitian, the eigenspaces corresponding to distinct eigenvalues λ_1 and λ_2 are orthogonal.

Proof: Let $Au_1 = \lambda_1 u_1$ and $Au_2 = \lambda_2 u_2$. Then

$$\lambda_1 u_1 \cdot u_2 = Au_1 \cdot u_2 = u_1 \cdot Au_2 = \lambda_2 u_1 \cdot u_2$$

Since $\lambda_1 \neq \lambda_2, u_1 \cdot u_2 = 0$, which proves the theorem.

The main theorem regarding Hermitian endomorphisms is the following.

Theorem 7.11: If A is a Hermitian endomorphism with (distinct) eigenvalues $\lambda_1, \lambda_2, \ldots, \lambda_N$, then V has the representation

$$V = V_1(\lambda_1) \oplus V(\lambda_2) \oplus \cdots \oplus V(\lambda_L) \tag{7.42}$$

where the eigenspaces $V(\lambda_k)$ are mutually orthogonal.

The proof of this theorem is a direct consequence of Theorems 7.35 and 7.36 and the remark following the proof of Theorem 7.6.

Corollary: (*Spectral Theorem*). If A is a Hermitian endomorphish with (distinct) eigenvalues $\lambda_1, \ldots, \lambda_L$ then

$$A = \sum_{j=1}^{N} \lambda_j P_j \tag{7.43}$$

where P_j is the perpendicular projection of V onto $V(\lambda_j)$, for $j = 1,\ldots,L$

Proof: By Theorem 7.11, A has a representation of the form (7.1). Let u be an arbitrary element of V, then, by (7.42),

$$u = u_1 + \ldots + u_L \tag{7.44}$$

where $u_j \in V(\lambda_j)$. By (7.3), (7.44), and (7.6)

$$A_j u = A_j u_j = A u_j = \lambda_j u_j$$

But $u_j = P_j u$; therefore

$$A_j = \lambda_j P_j \text{ (no sum)}$$

which, with (7.1) proves the corollary.

The reader is reminded that the L perpendicular projections satisfy the equations

$$\sum_{j=1}^{L} P_j = I \tag{7.45}$$

$$P_j^2 = P_j \tag{7.46}$$

$$P_j = P_j^* \tag{7.47}$$

and

$$P_j P_i = 0 \quad i \neq j \tag{7.48}$$

These equations follow from Theorems 5.26 and 5.27. Certain other endomorphisms also have a spectral representation of the form (7.43); however, the projections are not perpendicular ones and do not obey the condition (7.48).

Another Corollary of Theorem 7.11 is that if A is Hermitian, there exists an orthogonal basis for V consisting entirely of eigenvectors of A. This corollary is clear because each eigenspace is orthogonal to the other and within each eigenspace an orthogonal basis can be selected. (Theorem 4.3 ensures that an orthonormal basis for each eigenspace can be found.) With respect to this basis of eigenvectors, the matrix of A is clearly diagonal. Thus, the problem of finding a basis for V such that $M(A)$ is diagonal is solved for Hermitian endomorphisms.

If $f(A)$ is any polynomial in the Hermitian endomorphism, then (7.43), (7.46) and (7.48) can be used to show that

$$f(A) = \sum_{j=1}^{L} f(\lambda_j) P_j \tag{7.49}$$

where $f(\lambda)$ is the same polynomial except that the variable A is replaced by the scalar λ. For example, the polynomial P^2 has the representation

$$A^2 = \sum_{j=1}^{N} \lambda_j^2 P_j$$

In general, $f(A)$ is Hermitian if and only if $f(\lambda_j)$ is real for all $j = 1,\ldots,L$. If the eigenvalues of A are all non-negative, then we can extract Hermitian roots of A by the following rule:

$$A^{1/k} = \sum_{j=1}^{N} \lambda_j^{1/k} P_j \tag{7.50}$$

where $\lambda_j^{1/k} \geq 0$. Then we can verify easily that $(A^{1/k})^k = A$. If A has no zero eigenvalues, then

$$A^{-1} = \sum_{j=1}^{L} \frac{1}{\lambda_j} P_j \tag{7.51}$$

which is easily confirmed.

A Hermitian endomorphism A is defined to be

$$\begin{Bmatrix} \text{positive definite} \\ \text{positive semidefinite} \\ \text{negative semidefinite} \\ \text{negative definite} \end{Bmatrix} \text{ if } \mathrm{v} \cdot \mathrm{Av} \begin{Bmatrix} > 0 \\ \geq 0 \\ \leq 0 \\ < 0 \end{Bmatrix}$$

all non-zero v, It follows from (7.43) that

$$v \cdot Av = \sum_{j=1}^{L} \lambda_j v_j \cdot v_j \tag{7.52}$$

where

$$v = \sum_{j=1}^{L} v_j$$

Equation (7.52) implies the following important theorem.

Theorem 7.12: A Hermitian endomorphism A is

$$\begin{Bmatrix} \text{positive definite} \\ \text{positive semidefinite} \\ \text{negative semidefinite} \\ \text{negative definite} \end{Bmatrix}$$

if and only if every eigenvalue of A is

$$\begin{Bmatrix} > 0 \\ \geq 0 \\ \leq 0 \\ > 0 \end{Bmatrix}$$

As corollaries to Theorem 7.12 it follows that positive-definite and negative-definite Hermitian endomorphisms are regular [see (7.51)], and positive-definite and positive-semi-definite endomorphisms possess Hermitian roots.

Every complex number z has a polar representation in the form $z = re^{i\theta}$, where $r \geq 0$. It turns out that an endomorphism also has polar decompositions, and we shall deduce an important special case here.

Theorem 7.13: (*Polar Decomposition Theorem*). Every automorphism A has two unique multiplicative decompositions

$$A = RU \text{ and } A = VR \tag{7.53}$$

where R is unitary and U and V are Hermitian and positive definite.

Proof: For each automorphism A, A^*A is a positive-definite Hermitian endomorphism, and hence it has a spectral decomposition of the form

$$A^*A = \sum_{j=1}^{L} \lambda_j P_j \tag{7.54}$$

where $\lambda_j > 0$, $j = 1,\dots,L$. We define U by

$$U = \left(A^*A\right)^{1/2} = \sum_{j=1}^{L} \lambda_j^{1/2} P_j \tag{7.55}$$

Clearly U is Hermitian and positive definite. Since U is positive definite it is regular. We now define

$$R = AU^{-1} \tag{7.56}$$

By this formula R is regular and satisfies

$$R^*R = U^{-1}A^*AU^{-1} = U^{-1}U^2U^{-1} = I \tag{7.57}$$

Therefore, R is unitary. To prove the uniqueness, assume

$$RU = R_1U_1 \tag{7.58}$$

and we shall prove $R = R_1$ and $U = U_1$. From (7.58)

$$U^2 = UR^*RU = (RU)^* RU = (R_1U_1)\, R_1U_1 = U^2_1$$

Since the positive-definite Hermitian square root of U^2 is unique, we find $U = U_1$. Then (7.58) implies $R = R_1$. The decomposition (7.53) follows either by defining $V = RUR^*$, or equivalently, by defining $V = RUR^*$ and repeating the above argument.

A more general theorem to the last one is true even if A is not required to be regular. However, in this case R is not unique.

Before closing this section we mention again that if the scalar field is real, then a Hermitian endomorphism is symmetric, and a unitary endomorphism is orthogonal.

Exercises

1. Show that Theorem 7.10 remains valid if A is unitary or skew-Hermitian rather than Hermitian.

2. If $A \in L(V;V)$ is Hermitian, show that A is positive semi-definite if and only if the fundamental invariants of A are non-negative.
3. Given an endomorphism A, the *exponential* of A is an endomorphism $\exp A$ defined by the series

$$\exp A = \sum_{j=1}^{\infty} \frac{1}{j!} A^j \tag{7.59}$$

It is possible to show that this series converges in a definite sense for all A. Show that if $A \in L(V;V)$ is a Hermitian endomorphism given by the representation (7.43), then

$$\exp A = \sum_{j=1}^{L} e^{\lambda_j} P_j \tag{7.60}$$

This result shows that the series representation of $\exp A$ is consistent with (7.49).
4. Suppose that A is a positive-definite Hermitian endomorphism; give a definition for $\log A$ by power series and one by a formula similar to (7.49), then show that the two definitions are consistent. Further, prove that $\log \exp A = A$
5. For any $A \in L(V;V)$ show that A^*A and AA^* are Hermitian and positive semi-definite. Also, show that A is regular if and only if A^*A and AA^* are positive definite.
6. Show that $A^k = B^k$ where k is a positive integer, generally does not imply $A = B$ even if both A and B are Hermitian.

Illustrative Examples

In this section we shall illustrate certain of the results of the preceding section by working selected numerical examples. For simplicity, the basis of V shall be taken to be the orthonormal basis $\{i_1,\ldots,i_N\}$ introduced in (4.19). The vector equation (7.6) takes the component form

$$A_{kj}v_j = \lambda v_k \tag{7.61}$$

where

$$v = v_j i_j \tag{7.62}$$

and

$$Ai_j = A_{kj} i_k \tag{7.63}$$

Since the basis is orthonormal, we have written all indices as subscripts, and the summation convention is applied in the usual way.

Example 1: Consider a real three-dimensional vector space V. Let the matrix of an endomorphism $A \in L(V;V)$ be

$$M(A) = \begin{bmatrix} 1 & 1 & 0 \\ 1 & 2 & 1 \\ 0 & 1 & 1 \end{bmatrix} \tag{7.64}$$

Clearly A is symmetric and, thus, the theorems of the preceding section can be applied. By direct expansion of the determinant of $M(A-\lambda I)$ the characteristic polynomial is

$$f(\lambda) = (-\lambda)(1-\lambda)(3-\lambda) \tag{7.65}$$

Therefore, the three eigenvalues of A are distinct and are given by

$$\lambda_1 = 0,\ \lambda_2 = 1,\ \lambda_3 = 3 \tag{7.66}$$

The ordering of the eigenvalues is not important. Since the eigenvalues are distinct, their corresponding characteristic subspaces are one-dimensional. For definiteness, let $v^{(p)}$ be an eigenvector associated with λ_p. As usual, we can represent $v^{(p)}$ by

$$v^{(p)} = \text{v}_k^{(p)} i_k \tag{7.67}$$

Then (7.61), for p = 1, reduces to

$$\text{v}_1^{(1)} + \text{v}_2^{(1)} = 0,\ \text{v}_1^{(1)} + 2\text{v}_2^{(1)} + \text{v}_3^{(1)} = 0,\ \text{v}_2^{(1)} + \text{v}_3^{(1)} = 0 \tag{7.68}$$

The general solution of this linear system is

$$\text{v}_1^{(1)} = t,\ \text{v}_2^{(1)} = -t,\ \text{v}_3^{(1)} = t \tag{7.69}$$

for all $t \in R$. In particular, if $v^{(1)}$ is required to be a unit vector, then we can choose $t = \pm 1/\sqrt{3}$, where the choice of sign is arbitrary, say

$$\text{v}_1^{(1)} = 1/\sqrt{3},\ \text{v}_2^{(1)} = 1/\sqrt{3},\ \text{v}_3^{(1)} = 1/\sqrt{3} \tag{7.70}$$

So

$$v^{(1)} = \left(1/\sqrt{3}\right)i_1 - \left(1/\sqrt{3}\right)i_2 + \left(1/\sqrt{3}\right)i_3 \tag{7.71}$$

Likewise we find for $p = 2$

$$v^{(2)} = \left(1/\sqrt{2}\right)i_1 - \left(1/\sqrt{2}\right)i_3 \tag{7.72}$$

and for $p = 3$

$$v^{(3)} = \left(1/\sqrt{6}\right)i_1 + \left(2/\sqrt{6}\right)i_2 + \left(1/\sqrt{6}\right)i_3 \tag{7.73}$$

It is easy to check that $\{v^{(1)}, v^{(2)}, v^{(3)}\}$ is an orthonormal basis. By (7.66) and (7.43), A has the spectral decomposition

$$A = 0P_1 + 1P_2 + 3P_3 \tag{7.74}$$

where P_k is the perpendicular projection defined by

$$P_k v^{(k)} = v^{(k)},\ P_k v^{(j)} = 0,\ j \neq k \tag{7.75}$$

for k = 1,2,3. In component form relative to the original orthonormal basis $\{i_1, i_2, i_3\}$ these projections are given by

$$P_k i_j = \mathrm{v}_j^{(k)} \mathrm{v}_l^{(k)} i_l \text{ (no sum on } k) \tag{7.76}$$

This result follows from (7.75) and the transformation law (6.50) or directly from the representations

$$i_1 = \left(1/\sqrt{3}\right)v^{(1)} + \left(1/2\right)v^{(2)} + \left(1/\sqrt{6}\right)v^{(3)}$$

$$i_2 = -\left(1/\sqrt{3}\right)v^{(1)} + \left(2/\sqrt{6}\right)v^{(3)} \tag{7.77}$$

$$i_3 = \left(1/\sqrt{3}\right)v^{(1)} - \left(1/2\right)v^{(2)} + \left(1/\sqrt{6}\right)v^{(3)}$$

since the coefficient matrix of $\{i_1, i_2, i_3\}$ relative to $\{v^{(1)}, v^{(2)}, v^{(3)}\}$ is the transpose of that of $\{v^{(1)}, v^{(2)}, v^{(3)}\}$ relative to $\{i_1, i_2, i_3\}$.

There is a result, known as *Sylvester's Theorem*, which enables one to compute the projections directly. We shall not prove this theorem here, but we shall state the formula in the case when the eigenvalues are distinct. The result is

$$P_j = \frac{\coprod_{k=1;j\neq k}^{N}(\lambda_k I - A)}{\coprod_{k=1;j\neq k}^{N}(\lambda_k - \lambda_j)} \tag{7.78}$$

The advantage of this formula is that one does not need to know the eigenvectors in order to find the projections. With respect to an arbitrary basis, (7.78) yields

$$M(P_j) = \frac{\coprod_{k=1;j\neq k}^{N}(\lambda_k M(I) - M(A))}{\coprod_{k=1;j\neq k}^{N}(\lambda_k - \lambda_j)} \tag{7.79}$$

Example 2: To illustrate (7.79), let

$$M(A) = \begin{bmatrix} 2 & 2 \\ 2 & -1 \end{bmatrix} \tag{7.80}$$

with respect to an orthonormal basis. The eigenvalues of this matrix are easily found to be $\lambda_1 = -2, \lambda_2 = 3$. Then, (7.79) yields

$$M(p_1) = \left(\lambda_2 \begin{bmatrix} 1 & 0 \\ 0 & 1 \end{bmatrix} - \begin{bmatrix} 2 & 2 \\ 2 & -1 \end{bmatrix}\right) \Big/ (\lambda_2 - \lambda_1) = \frac{1}{5}\begin{bmatrix} 1 & -2 \\ -2 & 4 \end{bmatrix}$$

and

$$M(p_2) = \left(\lambda_1 \begin{bmatrix} 1 & 0 \\ 0 & 1 \end{bmatrix} - \begin{bmatrix} 2 & 2 \\ 2 & -1 \end{bmatrix}\right) \Big/ (\lambda_2 - \lambda_1) = \frac{1}{5}\begin{bmatrix} 4 & 2 \\ 2 & 1 \end{bmatrix}$$

The spectral theorem for this linear transformation yields the matrix equation

$$M(A) = -\frac{2}{5}\begin{bmatrix} 1 & -2 \\ -2 & 4 \end{bmatrix} + \frac{3}{5}\begin{bmatrix} 4 & 2 \\ 2 & 1 \end{bmatrix}$$

Exercises

1. The matrix of $A \in L(V;V)$ with respect to an orthonormal basis is

$$M(A) = \begin{bmatrix} 2 & 1 \\ 1 & 2 \end{bmatrix}$$

(a) Express $M(A)$ in its spectral form.

(b) Express $M(A^{-1})$ in its spectral form.

(c) Find the matrix of the square root of A.

2. The matrix of $A \in L(V;V)$ with respect to an orthonormal basis is

$$M(A) = \begin{bmatrix} 3 & 0 & 2 \\ 0 & 2 & 0 \\ 1 & 0 & 3 \end{bmatrix}$$

Find an orthonormal basis for V relative to which the matrix of A is diagonal.

3. If $A \in L(V;V)$ is defined by

$$Ai_1 = 2\sqrt{2i_1} - 2i_2$$

$$Ai_2 = \left(\frac{3}{2}\sqrt{2}+1\right)i_1 + \left(3-\frac{1}{2}\sqrt{2}\right)i_2$$

and

$$Ai_3 = i_3$$

where $\{i_1, i_2, i_3\}$ is an orthonormal basis for V, determine the linear transformations R, U, and V in the polar decomposition theorem.

Minimal Polynomial

We have remarked that for any given endomorphism A there exist some polynomials $f(t)$ such that

$$f(A) = 0 \tag{7.81}$$

For example, by the Cayley-Hamilton theorem, we can always choose, $f(t)$ to be the characteristic polynomial of A. Another obvious choice can be found by observing the fact that since $L(V;V)$ has dimension N^2, the set

$$\left\{A^0 = I, A^1, A^2, \ldots, A^{N^2}\right\}$$

is necessarily linearly dependent and, thus there exist scalars $\{\alpha_0, \alpha_1, \ldots, \alpha_{N^2}\}$, not all equal to zero, such that

$$\alpha_0 I + \alpha_1 A^1 + \cdots + \alpha_{N^2} A^{N^2} = 0 \tag{7.82}$$

For definiteness, let us denote the set of all polynomials f satisfying the condition (7.81) for a given A by the symbol $P(A)$. We shall now show that $P(A)$ has a very simple structure, called a *principal ideal.*

In general, an ideal I is a subset of an integral domain D such that the following two conditions are satisfied:

1. If f and g belong to I, so is their sum $f + g$.
2. If f belongs to I and h arbitrary element of D, then $fh = hf$ also belong to I.

Of course, D itself and the subset $\{0\}$ consisting in the zero element of D are obvious examples of ideals, and these are called *trivial ideals or improper ideals.* Another example of ideal is the subset $I \subset D$ consisting in all multiples of a particular element $g \in D$, namely

$$I = \{hg, h \in D\} \tag{7.83}$$

It is easy to verily that this subset satisfies the two conditions for an ideal. Ideals of the special form (7.83) are called *principal ideals.*

For the set $P(A)$ we choose the integral domain D to be the set of all polynomials with complex coefficients. (For real vector space, the coefficients are required to be real, of course.) Then it is obvious that $P(A)$ is an ideal in D, since if f and g satisfy the condition (7.81), so does their sum f + g and similarly if f satisfies (7.81) and h is an arbitrary polynomial, then

$$(hf)(A) = h(A)\, f(A) = h(A)0 = 0 \tag{7.84}$$

The fact that $P(A)$ is actually a principal ideal is a standard result in algebra, since we have the following theorem.

Theorem 7.14: Every ideal of the polynomial domain is a principal ideal.

Proof: We assume that the reader is similar with the operation of division for polynomials. If f and $g \neq 0$ are polynomials, we can divide f by g and obtain a remainder r having degree less than g, namely

$$r(t) = f(t) - h(t)g(t) \tag{7.85}$$

Now, to prove that $P(A)$ can be represented by the form (7.85), we choose a polynomial $g \neq 0$ having the lowest degree in $P(A)$. Then we claim that

$$P(A) = \{hg, h \in D\} \tag{7.86}$$

To see this, we must show that every $f \in P(A)$ can be devided through by g without a remainder. Suppose that the division of f by g yields a remainder r as shown in (7.85). Then since $P(A)$ is an ideal and since $f, g \in P(A)$, (7.85) shows that $r \in P(A)$ also. But since the degree of r is less than the degree of $g, r \in P(A)$ is possible if and only if $r = 0$. Thus $f = hg$, so the representation (7.86) is valid.

A corollary of the preceding theorem is the fact that the non-zero polynomial g having the lowest degree in $P(A)$ is unique to within an arbitrary non-zero multiple of a scalar. If we require the leading coefficient of g to be 1, then g becomes unique, and we call this particular polynomial g the *minimal polynomial* of the endomorphism A.

We pause here to give some examples of minimal polynomials.

Example 1: The minimal polynomial of the zero endomorphism 0 is the polynomial $f(t) = 1$ of zero degree, since by convention

$$f(0) = 10^{\circ} = 0 \tag{7.87}$$

In general, if $A \neq 0$, then the minimal polynomial of A is at least of degree 1, since in this case

$$1A^0 = 1I \neq 0 \tag{7.88}$$

Example 2: Let P be a non-trivial projection. Then the minimal polynomial g of P is

$$g(t) = t^2 - t \tag{7.89}$$

For, by the definition of a projection,

$$P^2 - P = P(P-I) = 0 \tag{7.90}$$

and since P is assumed to be non-trivial, the two lower degree divisors t and $t - 1$ no longer satisfy the condition (7.81) for P.

***Example* 3:** For the endomorphism C whose matrix is given by (7.30), the minimal polynomial g is

$$g(t) = t^N \tag{7.91}$$

since we have seen in that exercise that

$$C^N = 0 \text{ but } C^{N-1} \neq 0$$

For the proof of some theorems we need several other standard results in the algebra of polynomials. We summarise these results here.

Theorem 7.15: If f and g are polynomials, then there exists a greatest common divisor d which is a divisor (i.e., a factor) of f and g and is also a multiple of every common divisor of f and g.

***Proof*:** We define the ideal I in the polynomial domain D by

$$I \equiv \{hf + kg, h, k \in D\} \tag{7.92}$$

By Theorem 7.14, I is a principal ideal, and thus it has a representation

$$I \equiv \{hd, h \in D\} \tag{7.93}$$

We claim that d is a greatest common divisor of f and g. Clearly, d is a common divisor of f and g, since f and g are themselves members of I, so by (7.93) there exist h and k in D such that

$$f = hd, \; g = kd \tag{7.94}$$

On the other hand, since d is also a member of I, by (7.92) there exist also p and q in D such that

$$d = pf + qg \tag{7.95}$$

Therefore if c is any common divisor of f and g, say

$$f = ac,\ g = bc \tag{7.96}$$

then from (7.95)

$$d = (pa+qb)c \tag{7.97}$$

so d is a multiple of c. Thus, d is a greatest common divisor of f and g.

By the same argument as before, we see that the greatest common divisor d of f and g is unique to within a non-zero scalar factor. So we can render d unique by requiring its leading coefficient to be 1. Also, it is clear that the preceding theorem can be extended in an obvious way to more than two polynomials. If the greatest common divisor of $f_1,\dots,f_L$ is the zero degree polynomial, then $f_1,\dots,f_L$ are said to be *relatively prime*. Similarly $f_1,\dots,f_L$ are *pairwise prime* if each pair $f_i, f_j, i \neq j$, from $f_1,\dots,f_L$ is relatively prime.

Another important concept associated with the algebra of polynomials is the concept of the *least common multiple.*

Theorem 7.16: If f and g are polynomials, then there exists a *least common multiple* m which is a multiple of f and g and is a divisor of every common multiple of f and g.

The proof of this theorem is based on the same argument as the proof of the preceding theorem, so it is left as an exercise.

Exercises

1. If the eigenvalues of an endomorphism A are all single roots of the characteristic equation, show that the characteristic polynomial of A is also a minimal polynomial of A.
2. Prove Theorem 7.16.
3. If f and g are non-zero polynomials and if d is their greatest common divisor, show that then

   $m = fg/d$

   is their least common multiple and, conversely, if m is their least common multiplies, then

$d = fg/d$

is their greatest common divisor.

Spectral Decomposition for Arbitrary Endomorphisms

Not all endomorphims have matrices which are diagonal. However, for Hermitian endomorphisms, a decomposition of the endomorphism into a linear combination of projections is possible and is given by (7.43). In this section we shall consider the problem in general and we shall find decompositions which are, in some sense, closest to the simple decomposition (7.43) for endomorphisms in general.

We shall prove some preliminary theorems first. We have remarked that the null space $K(A)$ is always A-invariant. This result can be generalised to the following:

***Theorem** 7.17:* If f is any polynomial, then the null space $K(f(A))$ is A-invariant.

Proof: Since the multiplication of polynomials is a commutative operation, we have

$$Af(A) = f(A)A \tag{7.98}$$

Hence, if $v \in K(f(A))$, then

$$f(A)Av = Af(A)v = A0 = 0 \tag{7.99}$$

which shows that $Av \in K(f(A))$. Therefore, K(f(A)) is A-invariant.

Next we prove some theorems which describe the dependence of $K\,(f(A))$ on the choice of f.

***Theorem** 7.18:* If f is a multiple of g, say

$$f = hg \tag{7.100}$$

Then

$$K(f(A)) \supset K(g(A)) \tag{7.101}$$

Proof: This result is a general property of the null space. Indeed, for any endomorphisms B and C we always have

$$K(BC) \subset K(C) \tag{7.102}$$

So if we set $h(A) = B$ and $g(A) = C$, then (7.102) reduces to (7.101).

The preceding theorem does not imply that K $(g(A))$ is necessarily a proper subspace of K $(f(A))$, however. It is quite possible that the two subspaces, in fact, coincide. For example, if g and hence f both belong to $P(A)$, then $g(A) = f(A) = 0$, and thus

$$K(f(A)) = K(g(A)) = K(0) = V \tag{7.103}$$

However, if m is the minimal polynomial of A, and if f is a proper divisor of m (i.e., m is not a divisor of f) so that $f \notin P(A)$, then $K(f(A))$ is strictly a proper subspace of $V = K$ $(m(A))$. We can strengthen this result to the following.

Theorem 7.19: If f is a divisor (proper or improper) of the minimal polynomial m of A, and if g is a proper divisor of f, then K $(g(A))$ is strictly a proper subspace of K $(f(A))$.

Proof: By assumption there exists a polynomial h such that

$$m = hf \tag{7.104}$$

We set

$$k = hg \tag{7.105}$$

Then k is a proper divisor of m, since by assumption, g is a proper divisor of f. By the remark preceding the theorem, K $(k(A))$ is strictly a subspace of V, which is equal to K $(m(A))$. Thus, there exists a vector v such that

$$k(A)v = g(A)h(A)v \neq 0 \tag{7.106}$$

which implies that the vector $u = h(A)v$ does not belong to K $(g(A))$. On the other hand, from (7.104)

$$f(A)u == f(A)h(A)v = m(A)v = 0 \tag{7.107}$$

which implies that u belongs to K $(f(A))$. Thus, K $(g(A))$ is strictly a proper subspace of K $(f(A))$.

The next theorem shows the role of the greatest common divisor in terms of the null space.

Theorem 7.20: Let f and g be any polynomials, and suppose that d is their greatest common divisor. Then

$$K(d(A)) = K(f(A)) \cap K(g(A)) \qquad (7.108)$$

Obviously, this result can be generalised for more than two polynomials.

Proof: Since d is a common divisor of f and g, the inclusion

$$K(d(A)) \supset K(f(A)) \cap K(g(A)) \qquad (7.109)$$

follows readily from Theorem 7.99. To prove the reversed inclusion,

$$K(d(A)) \supset K(f(A)) \cap K(g(A)) \qquad (7.110)$$

recall that from (7.95) there exist polynomials p and q such that

$$d = pf + qg \qquad (7.111)$$

and thus

$$d(A) = p(A)f(A) + q(A)g(A) \qquad (7.112)$$

This equation means that if $v \in K(f(A)) \cap K(g(A))$, then

$$d(A)v = p(A)f(A)v + q(A)g(A)v$$

$$= p(A)\,0 + q(A)0 = 0 \qquad (7.113)$$

so that $v \in K(f(A))$. Therefore (7.110) is valid and hence (7.108).

A corollary of the preceding theorem is the fact that if f and g are relatively prime then

$$K(f(A)) \cap K(g(A)) = \{0\} \qquad (7.114)$$

since in this case the greatest common divisor of f and g is $d(t) = 1$, so that

$$K(d(A)) = K(A^0) = K(I) = \{0\} \qquad (7.115)$$

Here we have assumed $A \neq 0$ of course.

Next we consider the role of the least common multiple in terms of the null space.

Theorem 7.21: Let f and g be any polynomials, and suppose that l is their least common multiplier. Then

$$K(l(A)) = K(f(A)) + K(g(A)) \tag{7.116}$$

where the operation on the right-hand side of (7.116) is the sum of subspaces. Like the result (7.108), the result (7.116) can be generalised in an obvious way for more than two polynomials.

The proof of this theorem is based on the same argument as the proof of the preceding theorem, so it is left as an exercise. As in the preceding theorem, a corollary of this theorem is that if f and g are relatively prime (pairwise prime if there are more than two polynomials) then (7.116) can be strengthened to

$$K(l(A)) = K(f(A)) \oplus K(g(A)) \tag{7.117}$$

Again, we have assumed that $A \neq 0$. We leave the proof of (7.117) also as an exercise.

Having summarised the preliminary theorems, we are now ready to state the main theorem of this section.

Theorem 7.22: If m is the minimal polynomial of A which is factored into the form

$$m(t) = (t-\lambda_1)^{a_1} \cdots (t-\lambda_L)^{a_L} \equiv m_1(t) \cdots m_L(t) \tag{7.118}$$

where $\lambda_1, \ldots, \lambda_L$ are distinct and $a_1, \ldots, a_L$ are positive integers, then V has the representation

$$V = K(m_1(A)) \oplus \cdots \oplus K(m_L(A)) \tag{7.119}$$

Proof: Since $\lambda_1, \ldots, \lambda_L$ are distinct, the polynomials $m_1, \ldots, m_L$ are pairwise prime and their least common multiplier is m. Hence by (7.117) we have

$$k(m(A)) = K(m_1(A)) \oplus \cdots \oplus K(m_L(A)) \tag{7.120}$$

But since $m(A) = 0$, K(m(A)) = V, so that (7.119) holds.

Now from Theorem 7.17 we know that each subspace $K(m_i(A))$,.$i = 1, \ldots, L$ is A-invariant; then from (7.119) we see that A

is reduced by the subspaces $K(m_1(A)),\ldots,K(m_L(A))$. k{ ,{ })) .1 kfmL(6)). Therefore, the results of Theorem 7.1 can be applied. In particular, if we choose a basis for each $K(m_i(A))$ and form a basis for V by the union of these bases, then the matrix of A takes the form (7.5). The next theorem shows that, in some sense, the factorisation (7.118) gives also the minimal polynomials of the restrictions of A to the various A-invariant subspaces from the representation (7.119).

Theorem 7.23: Each factor m_k of m is the minimal polynomial of the restriction of A to the subspace $K(m_k(A))$. More generally, any product of factors, say $m_1 \cdots m_M$ is the minimal polynomial of the restriction of A to the corresponding subspace $k(m_1(A)) \oplus \cdots \oplus (m_M(A))$.

Proof: We prove the special case of one factor only, say m_1; the proof of the general case of several factors is similar and is left as an exercise. For definiteness, let $\overline{A}$ denote the restriction of A to $K(m_1(A)$. Then $m_1(\overline{A})$ is equal to the restriction of $m_1(A)$ to $K(m_1(A))$, and, thus $m_1(\overline{A}) = 0$, which means that $m_1 \in P(\overline{A})$. Now if g is any proper divisor of m_1, then $g(\overline{A}) \neq 0$; for otherwise, we would have $g(A)m_2(A)\cdots m_L(A)$, contradicting the fact that m is minimal for A. Therefore m_1 is minimal for $\overline{A}$ and the proof is complete.

The form (7.5), generally, is not diagonal. However, if the minimal polynomial (7.118) has simple roots only, i.e., the powers $a_1,\ldots,a_L$ are all equal to 1 , then

$$m_i(A) = A - \lambda_i I \tag{7.121}$$

for all i. In this case, the restriction of A to $K(m_i(A))$ coincides with $\lambda_i I$ on that subspace, namely

$$A_{v(\lambda_i)} = \lambda_i I \tag{7.122}$$

Here we have used the fact from (7.121), that

$$V(\lambda_i) = K(m_i(A)) \tag{7.123}$$

Then the form (7.5) reduces to the diagonal form [7.34(a)] or [7.34(b)].

The condition that m has simple roots only turns out to be necessary for the existence of a diagonal matrix for A also, as we shall now see in the following theorem.

Theorem 7.24: An endomorphism A has a diagonal matrix if and only if its minimal polynomial can be factored into distinct factors all of the first degree.

Proof: Sufficiency has already been proven. To prove necessity, assume that the matrix of A relative to some basis $\{e_1,\ldots,e_N\}$ has the form [7.34(a)]. Then the polynomial

$$m(t)=(t-\lambda_1)\cdots(t-\lambda_L) \tag{7.124}$$

is the minimal polynomial of A. Indeed, each basis vector e_i is contained in the null space $K(A-\lambda_k I)$ for one particular k. Consequently $e_i \in K(m(A))$ for all $i=1,\ldots,N$, and thus

$$K(m(A))=V \tag{7.125}$$

or equivalently

$$m(A)=0 \tag{7.126}$$

which implies that $m \in P(A)$. But since $\lambda_1,,,\lambda_{:L}$ are distinct, no proper divisor of m still belongs to $P(A)$. Therefore, m is the minimal polynomial of A.

As before, when the condition of the preceding theorem is satisfied, then we can define projections $P_1,\ldots,P_L$ by

$$R(P_i)=V(\lambda_i), \quad K(P_i)=\bigoplus_{\substack{j=1\\ j\neq i}}^{L} V(\lambda_j) \tag{7.127}$$

for all i = 1,...,L, and the diagonal form [7.34(a)] shows that A has the representation

$$A=\lambda_1 P_1+\cdots+\lambda_L P_L=\sum_{i=1}^{L}\lambda_i P_i \tag{7.128}$$

It should be noted, however, that in stating this result we have not made use of any inner product, so it is not meaningful to say whether or not the projections $P_1,\ldots,P_L$ are perpendicular; further, the eigenvalues $\lambda_1,\ldots,\lambda_L$ are generally complex numbers. In fact, the factorisation (7.118) for m in general is possible only if the scalar field is algebraically closed, such as the complex field used here. If the scalar field is the real field, we should define the factors $m_1,\ldots,m_L$ of m to be powers of irreducible polynomials, i.e., polynomials having no proper divisors. Then the decomposition (7.119) for V remains valid, since the argument of the proof is based entirely on the fact that the factors of m are pairwise prime.

Theorem 7.24 shows that in order to know whether or not A has a diagonal form, we must know the roots and their multiplicities in the minimal polynomial m of A. Now since the characteristic polynomial f of A belongs to $P(A)$, m is a divisor of f. Hence, the roots of m are always roots of f, The next theorem gives the converse of this result.

Theorem *7.25:* Each eigenvalue of A is a root of the minimal polynomial m of A and vice versa.

Proof: Sufficiency has already been proved. To prove necessity, let λ be an eigenvalue of A. Then we wish to show that the polynomial

$$g(t) = t - \lambda \tag{7.129}$$

is a divisor of m. Since g is of the first degree, if g is not a divisor of m, then m and g are relatively prime. By (7.114) and the fact that $m \in P(A)$, we have

$$\{0\} = K(m(A)) \cap K(g(A)) = V \cap K(g(A)) = K(g(A)) \tag{7.130}$$

But this is impossible, since $K(g(A))$, being the characteristic subspace corresponding to the eigenvalue λ, cannot be of zero dimension.

The preceding theorem justifies our using the same notations $\lambda_1,\ldots,\lambda_L$ for the (distinct) roots of the minimal polynomial m, as shown in (7.118). However, it should be noted that the root λ_i generally has a smaller multiplicity in m than in f, because m is a divisor of f. The characteristic polynomial f yields not only the (distinct) roots $\lambda_1,...,\lambda_L$ of m, it determines also the dimensions of their corresponding subspaces $K(m_1(A)),...,K(m_L(A))$ in the decomposition (7.119) This result is made explicit in the following theorem.

Theorem 7.26: Let d_k denote the algebraic multiplicity of the eigenvalue λ_k as before; i.e., d_k is the multiplicity of λ_k in f [cf. (7.19)]. Then we have

$$\dim K(m_k(A)) = d_k \tag{7.131}$$

Proof: We prove this result by induction. Clearly, it is valid for all A having a diagonal form, since in this case $m_k(A) = A - \lambda_k I$, so that $K(m_k(A))$ is the characteristic subspace corresponding to λ_k and its dimension is the geometric multiplicity as well as the algebraic multiplicity of λ_k. Now assuming that the result is valid for all A whose minimal polynomial has at most M multiple roots, where $M = 0$ is the starting induction hypothesis, we wish to show that the same holds for all A whose minimal polynomial has $M + 1$ multiple roots. To see this, we make use of the decomposition (7.120) and, for definiteness, we assume that λ_1 is a multiple root of m. We put

$$U = K(m_2(A)) \oplus \cdots \oplus K(m_L(A)) \tag{7.132}$$

Then U is A =invariant. From Theorem 7.23 we know that the minimal polynomial m_U of the restriction A_U is

$$m_U = m_2 \cdots m_L \tag{7.133}$$

which has at most M multiple roots. Hence by the induction hypothesis we have

$$\dim U = \dim K(m_2(A)) + \cdots + \dim K(m_L(A)) = d_2 + \cdots + d_L \tag{7.134}$$

But from (7.19) and (7.120) we have also

$$\begin{aligned} N &= d_1 + d_2 + \cdots + d_L \\ N &= \dim K(m_1(A)) + \dim K(m_2(A)) + \cdots + \dim K(m_L(A)) \end{aligned} \tag{7.135}$$

Comparing (7.135) with (7.134), we see that

$$d_1 = \dim K(m_1(A)) \tag{7.136}$$

Thus, the result (7.131) is valid for all A whose minimal polynomial has $M + 1$ multiple roots, and hence in general.

An immediate consequence of the preceding theorem is the following.

Theorem 7.27: Let b_k be the geometric multiplicity of λ_k, namely

$$d_k \equiv \dim V(\lambda_k) \equiv \dim K(g_k(A)) \equiv \dim K(A - \lambda_k I) \tag{7.137}$$

Then we have

$$1 \le b_k \le d_k - a_k + 1 \tag{7.138}$$

where d_k is the algebraic multiplicity of λ_k and a_k and is the multiplicity of λ_k in m, as shown in (7.118). Further, $b_k = d_k$ if and only if

$$K(g_k(A)) = K(g_k^2(A)) \tag{7.139}$$

Proof: If λ_k is a simple root of m, i.e., $a_k = 1$ and $g_k = m_k$, then from (7.131) and (7.137) we have $b_k = d_k$. On the other hand, if λ_k is a multiple root of m, i.e., $a_k > 1$ and $m_k = g_k^{a_k}$, then the polynomials $gk, g_k^2, \ldots, g_k^{(a_k - 1)}$ are proper divisors of m_k. Hence by Theorem 7.19.

$$V(\lambda_k) = K(g_k(A)) \subset K(g_k^2(A)) \subset \cdots \subset K(g_k^{(a_k - 1}(A)) \subset K(m_k(A)) \tag{7.140}$$

where the inclusions are strictly proper and the dimensions of the subspaces change by at least one in each inclusion. Thus (7.138) holds.

The second part of the theorem can be proved as follows: If λ_k is a simple root of m, then $m_k = g_k$, and, thus $K(g_k(\mathrm{A})) = K(g_k^2(A)) = V(\lambda_k)$. On the other hand, if λ_k is not a simple root of m, then m_k is at least of second degree. In this case g_k and g_k^2 are both divisors of m. But since g_k is also a proper divisor of g_k^2, by Theorem 7.19, $K(g_k(A))$ is strictly a proper subspace of $K(g_k^2(A))$, so that (7.139) cannot hold, and the proof is complete.

The preceding three theorems show that for each eigenvalue λ_k of A, generally there are two non-zero A -invariant subspaces, namely, the eigenspace $V(\lambda_k)$ and the subspace $K(m_k(A))$. For definiteness, let us call the latter subspace the characteristic subspace corresponding to λ_k and denote it by the more compact notation $U(\lambda_k)$. Then $V(\lambda_k)$ is a subspace of $U(\lambda_k)$ in general, and the two subspaces coincide if and only if λ_k is a simple root of m. Since λ_k is the only eigenvalue of the restriction of A to $U(\lambda_k)$, by the Cayley-Hamilton theorem we have also

$$U(\lambda_k) = K((A - \lambda_k I)^{d_k}) \tag{7.141}$$

where d_k is the algebraic multiplicity of λ_k, which is also the dimension of $U(\lambda_k)$. Thus we can determine the characteristic subspace directly from the characteristic polynomial of A by (7.141).

Now if we define P_k to be the projection on $U(\lambda_k)$ in the direction of the remaining $U(\lambda_j), j \neq K$, namely

$$R(P_k) = U(\lambda_k), K(P_k) = \bigoplus_{\substack{j=1 \\ j\neq k}}^{L} U(\lambda_j) \tag{7.142}$$

and we define B_k to be $A - \lambda_k P_k$ on $U(\lambda_k)$ and 0 on $U(\lambda_j), j \neq k$, then A has the spectral decomposition by a direct sum

$$A = \sum_{j=1}^{L} (\lambda_j P_j + B_j) \tag{7.143}$$

where

$$\left.\begin{aligned} P_j^2 &= P_j \\ P_j B_j &= B_j P_j = B_j \\ B_j^{a_j} &= 0,\ 1 \le a_j \le d_j \end{aligned}\right\} \quad j = 1, \ldots, L \tag{7.144}$$

$$\left.\begin{aligned} P_j P_k &= 0 \\ P_j B_k = B_k P_j &= 0 \\ B_j B_k &= 0, \end{aligned}\right\} \quad j \neq k,\ j,k = 1,\ldots,L \tag{7.145}$$

In general, an endomorphism B satisfying the condition

$$B^a = 0 \tag{7.146}$$

for some power a is called *nilpotent*. From (7.146) or from Theorem 7.25 the only eigenvalue of a nilpotent endomorphism is 0, and the lowest power a satisfying (7.146) is an integer $a, 1 \leq a \leq N$, such that t^N is the characteristic polynomial of B and t^a is the minimal polynomial of B. In view of (7.144) we see that each endomorphism B_j in the decomposition (7.143) is nilpotent and can be regarded also as a nilpotent endomorphism on $U(\lambda_j)$. In order to decompose A further from (7.143) we must determine a spectral decomposition for each B_j. This problem is solved in general as follows.

We have defined a nilcylic endomorphism C to be a nilpotent endomorphism such that

$$C^N = 0 \text{ but } C^{N-1} \neq 0 \tag{7.147}$$

Where N is the dimension of the underlying vector space V. For such an endomorphism we can find a cyclic basis $\{e_1,\ldots,e_N\}$ which satisfies the conditions

$$C^{N-1}e_1 = 0,\ C^{N-1}e_2 = e_1,\ldots,Ce_N = e_{N-1} \tag{7.148}$$

or, equivalently

$$C^{N-1}e_N = e_1,\ C^{N-2}e_2 = e_2,\ldots,Ce_N = e_{N-1} \tag{7.149}$$

so that the matrix of C takes the simple form (7.30). Indeed, we can choose e_N to be any vector such that $C^{N-1}e_N \neq 0$; then the set $\{e_1,\ldots,e_N\}$ defined by (7.149) is linearly independent and thus forms a cyclic basis for C. Nilcyclic endomorphisms constitute only a special class of nilpotent endomorphisms, but in some sense the former can be regarded as the building blocks for the latter. The result is made precise by the following theorem.

Theorem 7.28: Let B be a non-zero nilpotent endomorphism of V in general, say B satisfies the conditions

$$B^a = 0 \text{ but } B^{a-1} \neq 0 \tag{7.150}$$

for some integer a between 1 and N. Then there exists a direct sum decomposition for V:

$$V = V_1 \oplus \cdots \oplus V_M \tag{7.151}$$

and a corresponding direct sum decomposition for B:

$$B = B_1 + \cdots + B_M \tag{7.152}$$

such that each B_j is nilpotent and its restriction to V_j is nilcyclic. The subspaces $V_1,\ldots,V_M$ in the decomposition (7.151) are not unique, but their dimensions are unique and obey the following rules: The maximum of the dimension of $V_1,\ldots,V_M$ is equal to the integer a in (7.150); the number N_a of subspaces amonga $V_1,\ldots,V_M$ having dimension a is given by

$$N_a = N - \dim K\left(B^{a-1}\right) \tag{7.153}$$

More generally, the number N_b of subspaces among $V_1,\ldots,V_M$ having dimensions greater than or equal to b is given by

$$N_b = \dim K\left(B^b\right) - \dim K\left(B^{b-1}\right) \tag{7.154}$$

for all $b = 1,\ldots,a$. In particular, when $b = 1$, N_1 is equal to the integer M in (7.151), and reduces to

$$M = \dim K(B) \tag{7.155}$$

Proof: We prove the theorem by induction on the dimension of V. Clearly, the theorem is valid for one-dimensional space since a nilpotent endomorphism there is simply the zero endomorphism which is nilcylic. Assuming now the theorem is valid for vector spaces of dimension less than or equal to $N - 1$, we shall prove that the same is valid for vector spaces of dimension N.

Notice first if the integer a in (7.150) is equal to N, then B is nilcyclic and the assertion is trivially satisfied with $M = 1$, so we can assume that 1 < a < N. By (7.150), there exists a vector $e_a \in V$ such that $B^{a-1}e_a \neq 0$. As in (7.149) we define

$$B^{a-1}e_a = e_1, \ldots, Be_{a-1} \tag{7.156}$$

Then the set $\{e_1, \ldots, e_a\}$ is linearly independent. We put V_1 to be the subspace generated by $\{e_1, \ldots, e_a\}$. Then by definition $\dim V_1 = a$, and the restriction of B on V_1 is nilcyclic.

Now recall that for any subspace of a vector space we can define a factor space. As usual we denote the factor space of V over V_1 by V/V. From the result of (7.156), we have

$$\dim V/V_1 = N - a < N - 1 \tag{7.157}$$

Thus, we can apply the theorem to the factor space V/V_1. If $v \in V$, then $\overline{\text{v}}$ denotes the equivalence set of v in V/V_1. This notation means that the superimposed bar is the canonical projection from V to V/V_1. From (7.156) it is easy to see that V_1 is B-invariant. Hence, if u and v belong to the same equivalence set, so do Bu and Bv. Therefore, we can define an endomorphism $\overline{\text{B}}$ on the factor space V/V_1, by

$$\overline{B}\overline{v} = \overline{Bv} \tag{7.158}$$

for all $v \in V$ or equivalently for all $\overline{v} \in V/V_1$, Applying (7.157) repeatedly, we have also

$$\overline{B}^k\overline{v} = \overline{B^k v} \tag{7.159}$$

for all integers k. In particular, $\overline{B}$ is nilpotent and

$$\overline{B^a} = 0 \tag{7.160}$$

By the induction hypothesis we can then find a direct sum decomposition of the form

$$V/V_1 = U_1 \oplus \cdots \oplus U_p \tag{7.161}$$

for the factor space V/V_1 and a corresponding direct sum decomposition

$$\overline{B} = F_1 + \cdots F_p \tag{7.162}$$

for $\overline{B}$. In particular, there are cyclic bases in the subspaces $U_1, \ldots, U_p$ for the nilcyclic endomorphisms which are the restrictions of $F_1, \ldots, F_p$ to the corresponding subspaces. For definiteness, let $\{\overline{f_1}, \ldots, \overline{f_b}\}$ be a cyclic basis in U_1, say

$$\overline{B}^b \overline{f_b} = 0, \quad \overline{B}^{b-1} \overline{f_b} = \overline{f_1}, \ldots, \overline{Bf_b} = \overline{f}_{b-1} \tag{7.163}$$

From (7.160), is necessarily less than or equal to a.

From (7.163) and (7.159) we see that $\overline{B}^b \overline{f_b}$ belongs to V_1 and, thus can be expressed as a linear combination of $\{e_1, \ldots, e_a\}$, say

$$\overline{B}^b \overline{f_b} = \alpha_1 e_1 + \cdots + \alpha_a e_a = \left(\alpha_1 B^{a-1} + \cdots + \alpha_{a-1} B + \alpha_a I\right) e_a \tag{7.164}$$

Now there are two possibilities: $(i)\, \overline{B}^b \overline{f_b} = 0$ or $(ii)\, \overline{B}^b \overline{f_b} \neq 0$. In case (i) we define as before

$$B^{b-1} f_b = f_1, \ldots, Bf_b = f_{b-1} \tag{7.165}$$

Then $\{f_1, \ldots, f_b\}$ is a linearly independent set in V, and we put V_2 to be the subspace generated by $\{f_1, \ldots, f_b\}$, On the other hand, in case (ii) from (7.150) we see that b is strictly less than a; moreover, from (7.164) we have

$$0 = B^a f_b = \left(\alpha_{a-b+1} B^{a-1} + \cdots + \alpha_a B^{a-b}\right) e_a = \alpha_{a-b+1} e_1 + \cdots + \alpha_a e_b \tag{7.166}$$

which implies

$$= \alpha_{a-b+1} = \cdots = \alpha_a = 0 \tag{7.167}$$

or equivalently

$$B^b f_b = \left(\alpha_1 B^{a-1} + \cdots + \alpha_{a-b} B^b\right) e_a \tag{7.168}$$

Hence, we can choose another vector f_b' in the same equivalence set of f_b by

$$f_b' = f_b - \left(\alpha_1 B^{a-b-1} + \cdots + \alpha_{a-b} I\right) e_a = f_b - \alpha_1 e_{b+1} - \cdots - \alpha_{a-b} e_a \quad (7.169)$$

which now obeys the condition $B^b f_b' = 0$, and we can proceed in exactly the same way as in case (i). Thus, in any case every cyclic basis $\left\{\overline{f_1}, \ldots, \overline{f_b}\right\}$ for U_1 gives rise to a cyclic set $\left\{f_1, \ldots, f_b\right\}$ in V.

Applying this result to each one of the subspaces $U_1, \ldots, U_p$ we obtain cyclic sets $\left\{f_1, \ldots, f_b\right\}, \left\{g_1, \ldots, g_c\right\}, \ldots,$ and subspaces $V_2, \ldots, V_{p+1}$ generated by them in V. Now it is clear that the union of $\left\{e_1, \ldots, e_a\right\}, \left\{f_1, \ldots, f_b\right\}, \left\{g_1, \ldots, g_c\right\}, \ldots,$ form a basis of V since from (7.156), (7.157), and (7.161) there are precisely N vectors in the union; further, if we have

$$\alpha_1 e_1 + \cdots + \alpha_a e_a + \beta_1 f_1 + \cdots + \beta_b f_b + \gamma_1 g_1 + \cdots + \gamma_c g_c + \cdots = 0 \quad (7.170)$$

then taking the canonical projection to V/V yields

$$\beta_1 \overline{f_1} + \cdots + \beta_b \overline{f_b} + \gamma_1 \overline{g_1} + \cdots + \gamma_c \overline{g_c} + \cdots = 0$$

which implies

$$\beta_1 = \cdots = \beta_b = \gamma_1 = \cdots = \gamma_c = \cdots = 0$$

and substituting this result back into (7.170) yields

$$\alpha_1 = \cdots = \alpha_a = 0$$

Thus, V has a direct sum decomposition given by (7.151) with $M = p + 1$ and B has a corresponding decomposition given by (7.152) where $B_1, \ldots, B_M$ have the prescribed properties.

Now the only assertion yet to be proved is equation (7.154). This result follows from the general rule that for any nilcylic endomorphism C on a L-dimensional space we have

$$\dim K\left(C^k\right) - \dim\left(C^{k-1}\right) = \begin{cases} 1 & \text{for } 1 \leq k \leq L \\ 0 & \text{for } k > L \end{cases}$$

Applying this rule to the restriction of B_j to V for all $j = 1, \ldots, M$ and using the fact that the kernel of B^k is equal to the direct sum of the kernel of the restriction of $(B_j)^k$ for all $j = 1,\ldots,M$, prove easily that (7.154) holds. Thus, the proof is complete.

In general, we cannot expect the subspaces $V_1,\ldots,V_M$ in the decomposition (7.151) to be unique. Indeed, if there are two subspaces among $V_1,\ldots,V_M$ having the same dimension, say $\dim V_1 = \dim V_2 = a$, then we can decompose the direct sum $V_1 \oplus V_2$ in many other ways, e.g.,

$$V_1 \oplus V_2 = \bar{V}_1 \oplus \bar{V}_2 \tag{7.171}$$

and when we substitute (7.171) into (7.151) the new decomposition

$$V = \bar{V}_1 \oplus \bar{V}_2 \oplus V_3 \oplus \cdots \oplus V_N$$

possesses exactly the same properties as the original decomposition (7.151). For instance we can define V_1 and V_2 to be the subspaces generated by the linearly independent cyclic set $\{\tilde{e}_1,\ldots,\tilde{e}_a\}$ and $\{\tilde{f}_1,\ldots,\tilde{f}_a\}$, where we choose the starting vectors $\tilde{e}_a$ and $\tilde{f}_a$ by

$$\tilde{e}_a = \alpha e_a + \beta f_a, \ \tilde{f}_a = \gamma e_a + \delta f_a$$

provided that the coefficient matrix on the right hand side is non-singular.

If we apply the preceding theorem to the restriction of $A - \lambda_k I$ on U_k, we see that thek inequality (7.138) is the best possible one in general. Indeed, (7.138) becomes an equality if and only if U_k, has the decomposition

$$U_k = U_{k1} \oplus \ldots \oplus U_{kM} \tag{7.172}$$

where the dimensions of the subspaces $U_{k1},\ldots,U_{kM}$ are

$$\dim U_{k1} = a_k, \ \dim U_{k2} = \cdots = \dim U_{kM} = 1$$

If there are more than one subspaces among $U_{k1},\ldots,U_{kM}$ having dimension greater than one, then (7.138) is a strict inequality.

The matrix of the restriction of $A - \lambda_k I$ to U_k relative to the union of the cyclic basis for $U_{k1},\ldots,U_{kM}$ has the form

$$A_k = \begin{bmatrix} \boxed{A_{k1}} & & & & 0 \\ & \boxed{A_{k2}} & & & \\ & & \cdot & & \\ & & & \cdot & \\ & & & & \cdot \\ 0 & & & & \boxed{A_{kM}} \end{bmatrix} \tag{7.173}$$

where each submatrix A_{kj} in the diagonal of A_k has the form

$$A_{kj} = \begin{bmatrix} \lambda_k & 1 & & & 0 \\ & \lambda_k & 1 & & \\ & & \cdot & & \\ & & & \cdot & \\ & & & \cdot & 1 \\ 0 & & & & \lambda_k \end{bmatrix} \tag{7.174}$$

Substituting (7.173) into (7.5) yields the Jordan normal form for A:

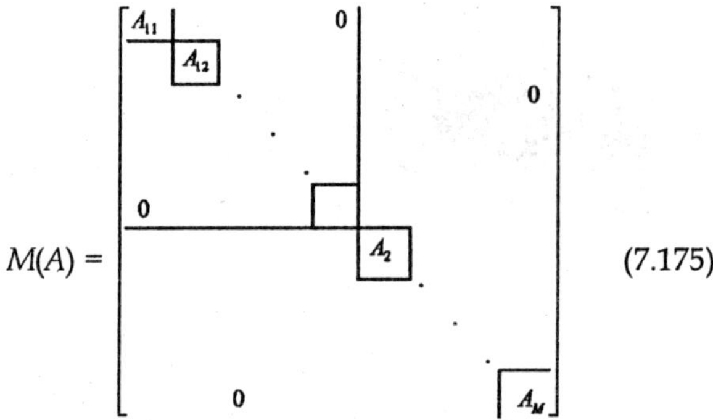

The Jordan normal form is an important result since it gives a geometric interpretation of an arbitrary endomorphism of a vector space. In general, we say that two endomorphisms A and A' are similar if the matrix of A relative to a basis is identical to the matrix of A' relative to another basis. From the transformation law (6.50), we see that A and A' are similar if and only if there exists a non-singular endomorphism T such that

$$A' = TAT^{-1} \tag{7.176}$$

Clearly, (7.176) defines an equivalence relation on $L(V;V)$. We call the equivalence sets relative to (7.176) the conjugate subsets of $L(V;V)$. Now for each $A \in L(V;V)$ the Jordan normal form of A is a particular matrix of A and is unique to within an arbitrary change of ordering of the various square blocks on the diagonal of the matrix. Hence, A and A' are similar if and only if they have the same Jordan normal form. Thus, the Jordan normal form characterises the conjugate subsets of $L(V;V)$.

Exercises

1. Prove Theorem 7.102.
2. Prove the general case of Theorem 7.104.
3. Let U be an unitary endomorphism of an inner product space V. Show that U has a diagonal form.
4. If V is a real inner product space, show that an orthogonal endomorphism Q in general does not have a diagonal form, but it has the spectral form

$$M(Q) = \begin{bmatrix} 1 & & & & & & & & & \\ & \ddots & & & & & & & & \\ & & 1 & & & & & 0 & & \\ & & & -1 & & & & & & \\ & & & & \ddots & & & & & \\ & & & & & -1 & & & & \\ & & & & & & \cos\theta_1 & -\sin\theta_1 & & \\ & & 0 & & & & \sin\theta_1 & \cos\theta_1 & & \\ & & & & & & & & \ddots & \\ & & & & & & & & \cos\theta_L & -\sin\theta_L \\ & & & & & & & & \sin\theta_L & \cos\theta_L \end{bmatrix}$$

   where the angles $\theta_1, \ldots, \theta_L$ may or may not be distinct.

5. Determine whether the endomorphism A whose matrix relative to a certain basis is

$$M(A) = \begin{bmatrix} 0 & 1 & 0 \\ 0 & 1 & 0 \\ 0 & -1 & 1 \end{bmatrix}$$

can have a diagonal matrix relative to another basis. Does the result depend on whether the scalar field is real or complex?

6. Determine the Jordan normal form for the endomorphism A whose matrix relative to a certain basis is

$$M(A) = \begin{bmatrix} 1 & 1 & 1 \\ 0 & 1 & 1 \\ 0 & 0 & 1 \end{bmatrix}$$

Tensor Algebra

The concept of a tensor is of major importance in applied mathematics. Virtually ever discipline in the physical sciences makes some use of tensors. Admittedly, one does not always need to spend a lot of time and effort to gain a computational facility with tensors as they are used in the applications. However, we take the position that a better understanding of tensors is obtained if we follow a systematic approach, using the language of finite-dimensional vector spaces. We begin with a brief discussion of linear functions on a vector space. Since in the applications the scalar field is usually the real field, from now on we shall consider real vector spaces only.

Linear Functions, the Dual Space

Let V be a real vector space of dimension N. We consider the space of linear functions $L\ (V; R)$ from V into the real numbers; R. By Theorem 5.11, $L\ (V; R)$ = dim $V = N$. Thus V and $L\ (V; R)$ are isomorphic. We call; $L\ (V; R)$ the *dual Space* of V, and we denote it by the special notation V^*. To distinguish elements of V from those of V^*, we shall call the former elements *vectors* and the latter elements *covectors*. However, these two names are strictly relative to each other. Since V^* is a N-dimensional vector space by itself, we can apply any result valid for a vector space in general to V^* as well as to V^*. In fact, we can even define a dual space $(V^*)^*$ for

V^* just as we define a dual space V^* for V. In this section V shall be a given N-dimensional space and V^* shall denote its dual space. As usual, we denote typical elements of V by u v, w,.... Then the typical elements of V^* are denoted by u^* v^*, w^*,... . However, it should be noted that the asterisk here is strictly a convenient notation, not a symbol for a function from V to V^*. Thus u^* is not related in any particular way to u. Also, for some covectors, such as those that constitute a dual basis to be defined shortly, this mutation becomes rather cumbersome. In such cases, the notation is simply abandoned. For instance, without fear of ambiguity we denote the null covector in the same notation as the null vector in V namely, 0, instead of 0^*.

If $v^* \in V^*$, then v^* is a linear function from V to R, *i.e.*,

$$v^* : V \to R$$

such that for any vectors u, v, $\in V$ and scalars $\alpha, \beta, \in R$

$$v^*(\alpha u + \beta v) = \alpha v^*(u) + \beta v^*(v)$$

Of course, the linear operations on the right hand side are those of R while those on the left hand side are linear operations in V. For a reason the will become apparent later, it is more convenient to denote the value of v^* at v by the notation $\langle v^*, v\rangle$. Then the bracket $\langle , \rangle$ operation can be viewed as a function

$$\langle , \rangle : V^* \times V \to R$$

It is easy to verify that this operation has the following properties:

(*i*) $\langle \alpha v^* + \beta u^*, v\rangle = \alpha\langle v^*, v\rangle + \beta\langle u^*, v\rangle$

(*ii*) $\langle v^*, \alpha u + \beta v\rangle = \alpha\langle v^*, u\rangle + \beta\langle v^*, v\rangle$ (8.1)

(*iii*) For any given $v, \langle v^*, v\rangle$ vanishes for all $v \in V$ if and only if $v^* = 0$.

(*iv*) Similarly, for any given $v, \langle v^*, v\rangle$ vanishes for all $v^* \in V^*$ if and only if $v = 0$.

The first two properties define $\langle , \rangle$ to be a *bilinear* operation on $V^* \times V$, and the last two properties define $\langle , \rangle$ to be a *definite* operation. These properties resemble the properties of an inner product, so that we call the operation $\langle , \rangle$ the *scalar product*. As we shall see, we can define many concepts associated with the scalar product similar to corresponding concepts associated with an inner product. The first example is the concept of the *dual basis,* which is the counterpart of the concept of the reciprocal basis.

If $\{e_1, \ldots e_N\}$ is a basis for V, we define it *dual basis* to be a basis $\{e_1, \ldots e_N\}$ for V^* such that

$$\left\langle e^j, e_i \right\rangle = \delta_i^j \tag{8.2}$$

for all $i, j = 1,\ldots,N$. The reader should compare this condition with the condition that defines the reciprocal basis. By exactly the same argument as before we can prove the following theorem.

Theorem 8.1: The dual basis relative to a given basis exists and it unique.

Notice that we have dropped the asterisk notation for the covector e^j in a dual basis; the superscript alone is enough to distinguish $\{e^1,\ldots e^N\}$ from $\{e_1,\ldots e_N\}$. However, it should be kept in mind that, unlike the reciprocal basis, the dual basis is a basis for V^*, not a basis for V. In particular, it makes no sense to require a basis be the same as its dual basis. This means the component form of a vector $v \in V$ relative to a basis $\{e_1, \ldots e_N\}$,

$$v = v^i e_i \tag{8.3}$$

must never be confused with the component form of a covector $V^* e V^*$ relative to dual basis,

$$v^* = v_i e^i \tag{8.4}$$

In order to emphasise the difference of these two component forms, we call v^i the *contravariant components* of v and v_i the *covariant components* of v^*. A vector has contravariant components only and a covector has covariant components only. The terminology for the components is not inconsistent with the same terminology defined earlier for an inner product space, since we have the following theorem.

Theorem 8.2: Given any inner product on V, there exists a unique isomorphism

$$G : V \to V^* \tag{8.5}$$

which is induced by the inner product in such a way that

$$\langle Gv, w \rangle = v.w, \qquad v, w \in V \tag{8.6}$$

Under this isomorphism the image of any orthonormal basis $\{i_1, \ldots i_N\}$ is the dual basis $\{i^1, \ldots i^N\}$, namely

$$Gi_k = i^k,\ k = 1, \ldots, N \tag{8.7}$$

and, more generally, if $\{e_1, \ldots e_N\}$ is an arbitrary basis, then the image of its reciprocal basis $\{\bar{e}^1, \ldots \bar{e}^N\}$, is the dual basis $\{e^1, \ldots e^N\}$ namely

$$G\bar{e}^k = e^k, \qquad k = 1, \ldots, N \tag{8.8}$$

Proof: Since we now consider only real vector spaces and real inner product spaces, the right-hand side of (8.6), clearly, is a linear function of w for each $v \in V$. Thus G is well defined by the condition (8.6). We must show that G is an isomorphism. The fact that G is a linear transformation is obvious, since the right-hand side of (8.6) is linear in v for each $w \in V$. Also, G is one-to-one because, from (8.6), if $Gu = Gv$, then $u.\ w = v.\ w$ for all w and thus $u = v$. Now since we already know that dim V = dim V^*, any one-to-one linear transformation from V to V^* is necessarily onto and hence an isomorphism. The proof of (8.8) is obvious, since by the definition of the reciprocal basis we have

$$\bar{e}^i \cdot e_j = \delta^i_j,\ i, j = 1, \ldots, N$$

and by the definition of the dual basis we have

$$\langle e^i, e_j \rangle = \delta^i_j, \qquad i, j = 1, \ldots, N$$

Comparing these definitions with (8.6), we obtain

$$\left\langle G\vec{e}^{j} = e_j \right\rangle = \left\langle e^i, e_j \right\rangle, \qquad i, j = 1,\ldots,N$$

which implies (8.8) because $\{e_1, \ldots e_N\}$ is a basis of V.

Because of this theorem, if a particular inner product is assigned on V, then we can, identify V with V^* by suppressing the notation for the isomorphisms G and G^{-1}. In other words, we regard a vector v also as a linear function on V:

$$\langle v, w \rangle = v.w \tag{8.9}$$

According to this rule the reciprocal basis is identified with the duel basis and the inner product becomes the scalar product. However, since a vector space can be equipped with many inner products, unless a particular inner product is chosen, we cannot identify V with V^* in general. In this section, we shall not assign any particular inner product in V, so V and V^* are different vector spaces.

We shall now derive some formulas which generalise the results of an inner product space to a vector space in general. First, if $v \in V$ and $v^* \in V^*$ are arbitrary, then their scalar products $\left\langle v^*, v \right\rangle$ can be computed in component form as follows: Choose a basis $\{e_i\}$ and its dual basis $\{e^i\}$ for V and V^*, respectively, so that we can express v and v^* in component form (8.3) and (8.4). Then from (8.1) and (8.2) we have

$$\left\langle v^*, v \right\rangle = \left\langle v_i e^i, v^j e_j \right\rangle = v_i v^j \left\langle e^i, e_j \right\rangle = v_i v^j \delta^i_j = v_i v^i \tag{8.10}$$

which generalises the formula (4.40). Applying (8.10) to $v^* = e^i$, we obtain

$$\left\langle e^i, v \right\rangle = v^i \tag{8.11}$$

which generalises the formula $(4.39)_1$; similarly applying (8.10) to $v = e_i$, we obtain

$$\left\langle v^*, e_i \right\rangle = v_i \tag{8.12}$$

which generalises the formula. $(4.39)_2$

Next recall that for inner product spaces V and U we define the adjoint A^* of a linear transformation A: $V \to U$ to be a linear transformation A^*: $U \to V$ such that the following condition is satisfied:

$$u.Av = A^* u.v, \qquad u \in U, \qquad v \in V$$

If we do not make use of any inner product, we simply replace this condition by

$$\langle u^*, Av \rangle = \langle A^* u^*, v \rangle, \quad u^* \in U^*, \quad v \in V \tag{8.13}$$

then A^* is a linear transformation from U^* to V^*,

$$A^*: U^* \to V^*$$

and is called the dual of A. By the same argument as before we can prove the following theorem.

Theorem 8.3: For every linear transformation $A: V \to U$ there exists a unique dual $A^*: U^* \to V^*$ satisfying the condition (8.13).

If we choose a basis $\{e_1, \ldots e_N\}$ for V and a basis $\{b_1, \ldots, b_M\}$ for U and express the linear transformation A by (5.41) and the linear transformation A^* by (5.42), where $\{b^\alpha\}$ and $\{e^k\}$ are now regarded as the dual bases of $\{b_\alpha\}$ and $\{e_k\}$ respectively, then (5.43) remains valid in the more general context, except hat we now have

$$A^{*}{}_{k}{}^{\alpha} = A^{\alpha}{}_{k} \tag{8.14}$$

since we no longer consider complex spaces. Of course, the formulas (5.44) and (5.45) are now replace by

$$\langle b^\alpha, Ae_k \rangle = A^\alpha{}_k \tag{8.15}$$

and

$$\langle A^* b^\alpha, e_k \rangle = A^{*}{}_{k}{}^{\alpha} \tag{8.16}$$

respectively.

For an inner product space the orthogonal complement of a subspace U of V is a subspace $U^\perp$ given by

$$U^\perp = \{v \mid u.v = 0 \text{ for all } u \in U\}$$

By the same token, if V is a vector space in general, then we define the *orthogonal complement* of U to be the subspace $U^{\perp}$ of V^* given by

$$U^{\perp} = \{V^* \mid \langle v^*, u \rangle = 0 \quad \text{for all } u \in U\} \tag{8.17}$$

In general if $v \in V$ and $v^* \in V^*$ are arbitrary, then v and v^* are said to be *orthogonal* to each $\langle v^*, v \rangle = 0$. We can prove the following theorem by the same argument used previously for inner product spaces.

Theorem 8.4: If U is a subspace of V, then

$$\dim U + \dim U^{\perp} = \dim V \tag{8.18}$$

However, V is no longer the direct sum of U and $U^{\perp}$ since $U^{\perp}$ is a subspace of V^*, not a subspace of V.

Using the same line of reasoning, we can generalise Theorem 5.42 to the following.

Theorem 8.5: If A: $V \rightarrow U$ is a linear transformation, then

$$K(A^*) = R(A)^{\perp} \tag{8.19}$$

and

$$K(A) = R(A^*)^{\perp} \tag{8.20}$$

Similarly, we can generalise Theorem 5.43 to the following.

Theorem 8.6: Any linear transformation A and its dual A^* have the same rank.

Finally, formulas for transferring from one basis to another basis can be generalised from inner product spaces to vector spaces in general. If $\{e_1, \dots e_N\}$ and $\{\hat{e}_1, \dots \hat{e}_N\}$ are bases for V, then as before we can express one basis in component form relative to another, as shown by (4.41) and (4.42). Now suppose that $\{e^1, \dots e^N\}$ and $\{\hat{e}^1, \dots \hat{e}^N\}$ are the dual bases of $\{e_1, \dots e_N\}$ and $\{\hat{e}_1, \dots \hat{e}_N\}$ respectively. Then it can be verified easily that

$$\hat{e}^q = \hat{T}^q_k e^k, \quad e^q = T^q_k \hat{e}^k \tag{8.21}$$

where T^q_k and $\hat{T}^q_k$ are defined by (4.41) and (4.42), i.e.,

$$e_k = \hat{T}^q_k \hat{e}_q, \quad \hat{e}_k = T^q_k e_q \tag{8.22}$$

From these relations if $v \in V$ and $v^* \in V^*$ have the component forms (8.3) and (8.4) relative to $\{e_i\}$ and $\{e^i\}$ and the component forms

$$v = \hat{u}^i \hat{e}_i \tag{8.23}$$

and

$$v^* = \hat{u}_i \hat{e}^i \tag{8.24}$$

relative to $\{\hat{e}_i\}$ and $\{\hat{e}^i\}$, respectively, then we have the following transformation laws:

$$\hat{u}_q = T^k_q u_k \tag{8.25}$$

$$\hat{u}^q = \hat{T}^q_k u^k \tag{8.26}$$

$$\hat{u}^k = \hat{T}^k_q \hat{u}^q \tag{8.27}$$

and

$$u_k = \hat{T}^q_k \hat{u}_q \tag{8.28}$$

which generalise the formulas (4.48)-(4.51), respectively.

Exercises

1. If $A : V \to U$ and $B : U \to W$ are linear transformations, show that

   $(BA)^* = A^* B^*$

2. If $A : V \to U$ and $B : V \to U$ are linear transformations, show that

   $(\alpha A + \beta B)^* = \alpha A^* + \beta B^*$

   and that

   $A^* = 0 \Leftrightarrow A = 0$

These two conditions mean that the operation of taking the dual is an isomorphism

$$*: L(V;U) \to L(U^*;V^*).$$

3. $A: V \to U$ is an isomorphism, show that $A^*: U^* \to V^*$ is also an isomorphism; moreover,

   $(A^{-1})^* = (A^*)^{-1}$

4. If V has the decomposition $V = U_1 \oplus U_2$ and if $P: V \to V$ is the projection of V on U_1 along U_2, show that V^* has the decomposition $V^* = U_1^{\perp} \oplus U_2^{\perp}$ and that $P^*: V^* \to V^*$ is the projection of V^* on $U_2^{\perp}$ along $U_1^{\perp}$.

5. If U_1 U_2 are subspaces of V, show that

   $(U_1 + U_2)^{\perp} = U_1^{\perp} \cap U_2^{\perp}$ and $(U_1 \cap U_2)^{\perp} = U_1^{\perp} + U_2^{\perp}$

6. Show that the linear transformation $G: V \to V^*$ defined in Theorem 8.2 obeys the condition

   $\langle Gv, w \rangle = \langle Gw, v \rangle$

7. Show that an inner product on V^* is induced by an inner product on V by the formula

   $$v^*.w^* \equiv G^{-1}v^*.G^{-1}w^* \tag{8.29}$$

   where G is the isomorphism defined in Theorem 8.2.

8. If $\{e_i\}$ is a basis for V and $\{e^i\}$ is its dual basis in V^*, show that $\{Ge_i\}$ is the reciprocal basis of $\{e^i\}$ with respect to the inner product on V^* defined by (8.29) in the preceding exercise.

Second Dual Space, Canonical Isomorphisms

We defined the dual space V^* of any vector space V to be the space of linear functions $L(V; R)$ from V to R. By the same procedure we can define the dual space $(V^*)^*$ of V^* by

$$(V^*)^* = L(V^*; R) = L\,(L(V; R); R) \tag{8.30}$$

For simplicity let us denote this space by V^{**}, called the *second dual space* of V. Of course, the dimension of V^{**} is the same as that of V, namely

$$\dim V = \dim V^{*} = \dim V^{**} \tag{8.31}$$

Using the system of notation, we write a typical element of V^{**} by v^{**}. Then v^{**} is a linear function on V^{*}

$$v^{**} : V^{*} \to R$$

Further, for each $v^{*} \in V^{*}$ we denote the value of V^{**} at v^{*} by $\langle v^{**}, v^{*} \rangle$. The $\langle , \rangle$ operation is now a mapping from $V^{**} \times V^{*}$ to R.

Unlike the dual space V^{*}, the second dual space V^{**} can always be identified as V without using any inner product. The isomorphism

$$J : V \to V^{**} \tag{8.32}$$

is defined by the condition

$$\langle Jv, v^{*} \rangle = \langle v^{*}, v \rangle, \qquad v \in V, \quad v^{*} \in V^{*} \tag{8.33}$$

Clearly, J is well defined by (8.33) since for each $v \in V$ the right-hand side of (8.33) is a linear function of v^{*}. To see that J is an isomorphism, we notice first that J is a linear transformation, because for each $v^{*} \in V^{*}$ the right-hand side is linear in v. Now J also one-to-one, since if $Jv = Ju$, then (8.33) implies that

$$\langle v^{*}, v \rangle = \langle v^{*}, u \rangle, \quad v^{*} \in V^{*} \tag{8.34}$$

which then implies $u = v$ From (8.31), we conclude that the one-to-one linear transformation J is onto, and thus J is an isomorphism. We summarise this result in the following theorem.

Theorem 8.7: There exists a unique isomorphism J from V to V^{**} satisfying the condition (8.33).

Since the isomorphism J is defined without using any structure in addition to the vector space structure on V, its notation can often be suppressed without any ambiguity. We shall, adopt such a convention here and identify any $v \in V$ as a linear function on V^{*}

$$v : V^{*} \to R$$

by the condition that defines J, namely

$$\langle v, v^* \rangle = \langle v^*, v \rangle, \text{ for all } v^* \in V^* \qquad (8.35)$$

In doing so, we allow the same symbol v to represent two different objects: an element of the vector space V and a linear function on the vector space V^*, and the two objects are related to each other through the condition (8.35).

To distinguish an isomorphism such as J, whose notation may be suppressed without, causing any ambiguity, from an isomorphism such as G defined by (8.6), whose notation may, not be suppressed, because there are many isomorphisms of similar nature, we call the former isomorphism a *canonical or natural isomorphism*. Whether or not an isomorphism is canonical is usually determined by a convention, not by any axioms.

A general rule for choosing a canonical isomorphism is that the isomorphism must be defined without using any additional structure other than the basic structure already assigned to the underlying spaces; further, by suppressing the notation of the canonical isomorphism no ambiguity is likely to arise. Hence the choice of a canonical isomorphism depends on the basic structure of the vector spaces. If we deal with inner product spaces equipped with particular inner products, the isomorphism G can safely be regarded as canonical, and by choosing G to be canonical, we can achieve much economy in writing. On the other hand, if we consider vector spaces without any preassigned inner product, then we cannot make all possible isomorphisms G canonical, otherwise the notation becomes ambiguous.

It should be noticed that not every isomorphism whose definition depends only on the basis structure of the underlying space can be made canonical. For example, the operation of taking the dual:

$$* : L(V;U) \to L(U^*;V^*)$$

is defined by using the vector space structure of V and U only. However, by suppressing the notation $*$, we encounter immediately much ambiguity, especially when U is equal to V. Surely, we do not wish to make every endomorphism $A: V \to V$ self-adjoint! Another example will illustrate the point even clearer. The operation of taking the opposite vector of any vector is an isomorphism

$$-: V \to V$$

which is defined by using the vector space structure alone. Evidently, we cannot suppress the minus sign without any ambiguity.

To test whether the isomorphism J can be made a canonical one without any ambiguity, we consider the effect of this choice on the notations for the dual basis and the dual of a linear transformation. Of course we wish to have $\{e_i\}$, when considered as a basis for V^{**}, to be the dual basis of $\{e^i\}$, and A, when considered as a linear transformation from, V^{**} to U^{**}, to be the dual of A^*. These results are indeed correct and they are contained in the following.

Theorem 8.8: Given any basis $\{e_i\}$ for V, then the dual basis of its dual basis $\{e^i\}$ is $\{Je_i\}$.

Proof: This result is more or less obvious. By definition, the basis $\{e_i\}$ and its dual basis $\{e^i\}$ are related by

$$\left\langle e^i, e_j \right\rangle = \delta^i_j$$

for, i, j = 1, ...N. From (8.33) we have

$$\left\langle Je_j, e^i \right\rangle = \left\langle e^i, e_j \right\rangle$$

Comparing the preceding two equations, we see that

$$\left\langle Je_j, e^i \right\rangle = \delta^i_j$$

which means that $\{Je_1, \ldots Je_N\}$ is the dual basis of $\{e^1, \ldots e^N\}$.

Because of this theorem, after suppressing the notation for J we say that $\{e_i\}$ and $\{e^i\}$ are dual relative to each other. Ther next theorem shows that the relation holds between A and A^*.

Theorem 8.9: Given any linear transformation $A: V \to U$, the dual of its dual A^* is $J_U A J_V^{-1}$. Here J_V denotes the isomorphism from V to V^{**} defined by (8.33) and J_U denotes the isomorphism from U to U^{**} defined by a similar condition.

Proof: By definition A and A^* are related by (8.13)

$$\langle u^*, Av\rangle = \langle A^* u^*, v\rangle$$

for all $u^* \in U^*, v \in V$. From (8.33) we have

$$\langle u^*, Av\rangle = \langle J_U Av, u^*\rangle, \ \langle A^* u^*, v\rangle = \langle J_V v, A^* u^*\rangle$$

Comparing the preceding three equations, we see that

$$\langle J_U A J_V^{-1}(J_V v), u^*\rangle = \langle J_V v, A^* u^*\rangle$$

Since J_V is an isomorphism, we can rewrite the last equation as

$$\langle J_U A J_V^{-1} v^{**}, u^*\rangle = \langle v^{**}, A^* u^*\rangle$$

Because v^{**} ÎV^{**} and $u^* \in U^*$ are arbitrary, it follows that $J_U A J_V^{-1}$ is the dual of A^*. So if we suppress the notations for J_U and J_V, then A and A^* are the duals relative to each other.

A similar result exists for the operation of taking the orthogonal complement of a subspace; we have the following result.

Theorem 8.10: Given any subspace U of V, the orthogonal complement of its orthogonal complement $U^{\perp}$ is J (U).

We leave the proof of this theorem as an exercise. Because of this theorem we say that U and $U^{\perp}$ are orthogonal to each other. The use of canonical isomorphisms, like the summation convention, is an important device to achieve economy in writing. We shall make use of this device whenever possible, so the reader should be prepared to allow one symbol to represent two or more different objects.

The last three theorems show clearly the advantage of making J a canonical isomorphism, so from now on we shall suppress the symbol for J. In general, if an isomorphism from a vector space V to a vector space U is chosen to be canonical, then we write

$$V \cong U \tag{8.36}$$

In particular, we have

$$V \cong V^{**} \tag{8.37}$$

Exercises

1. Prove theorem 8.10.
2. Show that by making J a canonical isomorphism essentially we have identified V with V^{**}, V^{****},..., and V^* with V^{***}, V^{*****},... . So a symbol v or v^*, in fact, represents, infinitely many objects.

Multilinear Functions, Tensors

The concept of a linear function and the concept of a bilinear function can be generalised in an obvious way to *multilinear functions.* In general if $V_1,..., V_s$, is a collection of vector spaces, then a *s-linear function* is a function

$$A: V_1 \times ... \times V_s \to R \tag{8.38}$$

that is linear in each of its variables while the other variables are held constant. If the vector spaces $V_1,...,V_s$, are the vector space V or its dual space V^*, then A is called a *tensor* on V. More specifically, a *tensor of order* (p, q) on V, where p and q are positive integers, is a $(p+q)$- linear function

$$\underbrace{V^* \times ... \times V^*}_{p\,\text{times}} \times \underbrace{V \times ... \times V}_{q\,\text{times}} \to R \tag{8.39}$$

We shall extend this definition to the case $p = q=0$ and define a tensor of order (0,0) to be a scalar in R. A tensor of order $(p,0)$ is a pure *contravariant* tensor of order p and a tensor of order $(0, q)$ is a pure *covariant* tensor of order q. In particular, a vector $v \in V$ is a pure contravariant tensor of order one. This terminology, of course, is defined relative to a given vector space V. If a tensor is not a pure contravariant tensor or a pure covariant tensor, then it is a mixed tensor, and for a mixed tensor of order (p, q), p is the *contravariant* order and q is the *covariant* order.

For definiteness, we denote the set of all tensors of order (p, q) on V by the symbol $T_q^p(V)$ However, the set of pure contravariant tensors of order p shall be denoted simply by $T^p(V)$ and the set of pure covariant tensors of order q shall be denoted simply $T_q(V)$. Of course, tensors of order (0,0) form the set R and

$$T_1(V) = V, \; T_1(V) = V^* \tag{8.40}$$

Here we have made use of the identification of V and V^{**}.

We shall now give some examples of tensors.

Example 1: If A is an endomorphism of V, then we define a function $\hat{A}: V^* \times V \to R$ by

$$\hat{A}(v^*, v) \equiv \left\langle v^*, Av \right\rangle \tag{8.41}$$

for all $v^* \in V^*$ and $v \in V$. Clearly $\hat{A}$ is bilinear and thus $\hat{A} \in T_1^{\,1}(V)$. As we shall see later, it is possible to establish a canonical isomorphism from $L(V,V)$ to $T_1^{\,1}(V)$ in such a way that the endomorphism A is identified with the bilinear function $\hat{A}$. Then the same symbol A shall represent two objects, namely, an endomorphism of V and a bilinear function of $V^* \times V$. Then (8.41) becomes simply

$$A(v^*, v) \equiv \left\langle v^*, Av \right\rangle \tag{8.42}$$

Under this canonical isomorphism, the identity automorphism of V is identified with the scalar product,

$$I(v^*, v) = \left\langle v^*, Iv \right\rangle = \left\langle v^*, v \right\rangle \tag{8.43}$$

Example 2: If v is a ventor in V and v^* is a covector in V^*, then we define a function

$$v \otimes v^* : V \times V^* \to R \tag{8.44}$$

by

$$v \otimes v^*(u^*, u) \equiv \left\langle u^*, u \right\rangle \left\langle v^*, u \right\rangle \tag{8.45}$$

for all $u^* \in V^*, u \in V$. Clearly, $v \otimes v^*$ is a bilinear function, so $v \otimes v^* \in T_1^{\,1}(V)$. If we make use of the canonical isomorphism to be established between $T_1^{\,1}(V)$ and $L(V;V)$, the tensor $v\otimes v^*$ corresponds to an endomorphism of V such that

$$v\otimes v^*(u^*,u)=\left\langle u^*,v\otimes v^*u\right\rangle$$

or equivalently

$$\left\langle u^*,v\right\rangle\left\langle v^*,u\right\rangle=\left\langle u^*,v\otimes v^*u\right\rangle$$

for all $u^*\in V^*, u\in V$. By the bilinearity of the scalar product, the last equation can be rewritten in the form

$$\left\langle u^*,\left\langle v^*,u\right\rangle v\right\rangle=\left\langle u^*,v\otimes v^*u\right\rangle$$

Then by the definiteness of the scalar product, we have

$$\left\langle v^*,u\right\rangle v=v\otimes v^*u, \qquad \text{for all } u\in V \tag{8.46}$$

which defines $v\otimes v^*$ as an endomorphism of V. The tensor or the endomorphism $v\otimes v^*$ is called the *tensor product* of v and v^*.

Clearly, the tensor product can be defined for arbitrary number of vectors and covectors. Let $v_1, \ldots, v_p$ be vectors in V and $v^1, \ldots, v^q$ be covectors in V^*. Then we define a function

$$v_1\otimes\cdots\otimes v_p\otimes v^1\otimes\cdots\otimes v^q : \underbrace{V^*\times\cdots\times V^*}_{p\,\text{times}}\times\underbrace{V\times\cdots\times V}_{q\,\text{times}}\to R$$

by

$$\begin{aligned} v_1\otimes\cdots\otimes v_p\otimes v^1\otimes\cdots\otimes v^q(u^1,\ldots,u^p,u_1,\ldots,u_q)& \\ \equiv\left\langle u^1,v_1\right\rangle\ldots\left\langle u^p,v_p\right\rangle\left\langle v^1,u_1\right\rangle\ldots\left\langle v^q,u_q\right\rangle& \end{aligned} \tag{8.47}$$

for all $u_1,\ldots,u_q\in V$ and $u^1,\ldots,u^q\in V^*$. Clearly this function is $(p+q)$ - linear, so that $v_1\otimes\cdots\otimes v_p\otimes v^1\otimes\cdots\otimes v^q\in T_q^p(V)$, is called the *tensor product* of $v_1,\ldots,v_p$ and $v^1,\ldots,v^q$.

Having seen some examples of tensors on V, we turn now to the structure of the set, $T_q^p(V)$. We claim that $T_q^p(V)$ has the structure of a vector space and the dimension of $T_q^p(V)$ is equal to $N^{(p+q)}$, where N is the dimension of V. To make $T_q^p(V)$ a vector space, we define the operation of addition of any $A, B \in T_q^p(V)$ and the scalar multiplication of A by $\alpha \in R$ by

$$(A+B)\ (v^1,\ldots,v^p,\ v^1,\ldots v_q)$$

$$\equiv \text{A}\ (v^1,\ldots,v^p,\ v_1,\ldots v_q) + \text{B}\ (v^1,\ldots,v^p,\ v_1,\ldots v_q) \tag{8.48}$$

and

$$(\alpha A)(v^1,\ldots v^p, v_1,\ldots v_q) \equiv \alpha A(v^1,\ldots,v^p, v_1,\ldots v_q) \tag{8.49}$$

respectively, for all $v^1,\ldots,v^p \in V^*$ and $v_1,\ldots,v_p \in V$. We leave as an exercise to the reader the proof of the following theorem.

Theorem 8.11: $T_q^p(V)$ is a vector space with respect to the operations of addition and scalar multiplication defined by (8.48) and (8.49). The null element of $T_q^p(V)$, of course, is the zero tensor 0:

$$0(v^1,\ldots,v_p,\ v^1,\ldots,v_q)=0 \tag{8.50}$$

for all $v^1,\ldots,v^p \in V^*$ and $v_1,\ldots,v_q \in V$.

Next, we determine the dimension of the vector space $T_q^p(V)$ by introducing the concept of a *product basis.*

Theorem 8.12: Let $\{e_i\}$ and $\{e^i\}$ be dual bases for V and V^*. Then the set of tensor products

$$\left\{e_{i_1} \otimes \cdots \otimes e_{i_p} \otimes e^{j_1} \otimes \cdots \otimes e^{i_q}, i_1,\ldots,i_p,\ldots,j_1,\ldots,j_q = 1,\ldots,N\right\} \tag{8.51}$$

forms a basis for $T_q^p(V)$, called the *product basis.* In particular,

$$\dim\ T_q^p(V) = N^{(p+q)} \tag{8.52}$$

Proof: We shall prove that the set of tensor products (8.51) is a linearly independent generating set for $T_q^p(V)$. To prove that the set (8.51) is linearly independent, let

$$A^{i_1\ldots,i_p}{}_{j_1,\ldots j_q} e_{i_1} \otimes \cdots \otimes e_{i_p} \otimes e^{j_i} \otimes \cdots \otimes e^{i_q} = 0 \tag{8.53}$$

where the right-hand side is the zero tensor given by (8.50). Then from (8.47) we have

$$\begin{aligned} 0 &= A^{i_1 \dots i_p}{}_{j_1 \dots j_q} e_{i_1} \otimes \cdots \otimes e_{i_p} \otimes e^{j_q} \otimes \cdots \otimes e^{j_q} (e^{k_1}, \dots, e^{k_p}, e_{1_1}, \dots e_{1_q}) \\ &= A^{i_1 \dots i_p}{}_{j_1 \dots j_q} \left\langle e^{k_1}, e_{i_1} \right\rangle \cdots \left\langle e^{k_p}, e_{i_p} \right\rangle \left\langle e^{j_1}, e_{l_1} \right\rangle \cdots \left\langle e^{j_q}, e_{l_q} \right\rangle \\ &= A^{i_1 \dots i_p}{}_{j_1 \dots j_q} \delta^{k_1}_{i_1} \dots \delta^{k_p}_{i_p} \delta^{j_1}_{l_1} \dots \delta^{j_q}_{l_q} = A^{k_1 \dots k_p}{}_{l_1 \dots l_q} \end{aligned} \tag{8.54}$$

which shows that the set (8.51) is linearly independent. Next, we show that every tensor $A \in T^p_q(V)$ can be expressed as a linear combination of the set (8.51). We define $N^{(p+q)}$ scalars $\left\{ A^{i_1 \dots i_p}{}_{j_1 \dots j_q}, i_1 \dots i_p, j_1 \dots j_q = 1, \dots, N \right\}$ by

$$A^{i_1 \dots i_p}{}_{j_1 \dots j_q} = A(e^{i_1}, \dots, e^{i_p}, e_{j_1}, \dots, e^{j_q}) \tag{8.55}$$

Now we reverse the steps of equation (8.54) and obtain

$$\begin{aligned} A^{i_1 \dots i_p}{}_{j_1 \dots j_q} e_{i_1} \otimes \cdots \otimes e_{i_p} \otimes e^{j_1} \otimes \cdots \otimes e^{j_q} (e^{k_1}, \dots, e^{k_p}, e_{l_1}, \dots, e_{l_q}) \\ = A(e^{k_1}, \dots, e^{k_p}, e_{l_1}, \dots e_{l_q}) \end{aligned} \tag{8.56}$$

for all $k_1, \dots, k_p, l_1, \dots, l_q = 1, \dots, N$. Since A is multilinear and since $\{e_i\}$ and $\{e^i\}$ are dual bases for V and V^*, the condition (8.56) implies that

$$\begin{aligned} A^{i_1 \dots i_p}{}_{j_1 \dots j_q} e_{i_1} \otimes \cdots \otimes e_{i_p} \otimes e^{j_1} \otimes \cdots \otimes e^{j_q} (v^1, \dots, v^p, v_1, \dots, v_q) \\ = A(v^1, \dots, v^p, v_1, \dots v_q) \end{aligned}$$

for all $v_1, \dots, v_q \in V$ and $v^1, \dots, v^p \in V^*$ and thus

$$A = A^{i_1 \dots i_p}{}_{j_1 \dots j_q} e_{i_1} \otimes \cdots \otimes e_{i_p} \otimes e^{j_1} \otimes \cdots \otimes e^{j_q} \tag{8.57}$$

Now from Theorem 4.13 we conclude that the set (8.51) is a basis for $T^p_q(V)$.

Having determined the dimension of $T^p_q(V)$, we can now prove that $T^1_1(V)$ is isomorphic to L $(V; V)$, a result mentioned in Example 1. As before we define the operation

$$\hat{}: L(V;V) \to T_1^1(V)$$

by the condition (8.41). Since from (5.15) and (8.52), the vector space $L(V;V)$ and $T_1^1(V)$ are of the same dimension, namely N^2, it suffices to show that the operation ^ is one-to-one. Indeed, let $\hat{A} = 0$. Then from (8.50) and (8.41) we have

$$\langle v^*, Av \rangle = 0$$

for all $v^* \in V^*$ and all $v \in V$. Now since the scalar product is definite, this condition implies

$$Av = 0$$

for all $v \in V$, and thus $A = 0$. Consequently the operation ^ is an isomorphism.

As remarked in Example 1, we shall suppress the notation ^ by regarding the isomorphism it represents as a canonical one. Therefore, we can replace the formula (8.41) by the formula (8.42).

Now returning to the space $T_q^p(V)$ in general, we see that a corollary of Theorem 8.12 is the simple fact that the set of all tensor products of the form (8.47) is a generating set for $T_q^p(V)$. We define a tensor that can be represented as a tensor product to be a *simple tensor*. It should be noted, however, that such a representation for a simple tensor is not unique. Indeed, from (8.45), we have

$$v \otimes v^* = (2v) \otimes \left(\frac{1}{2} v^*\right)$$

for any $v \in V$ and $v^* \in V^*$ since

$$\begin{aligned}(2v) \otimes \left(\frac{1}{2} v^*\right)(u^*, u) &= \langle u^*, 2v \rangle \left\langle \frac{1}{2} v^*, u \right\rangle \\ &= \langle u^*, v \rangle \langle v^*, u \rangle = v \otimes v^*(u^*, u)\end{aligned}$$

for all u^* and u. In general, the tensor product, as defined by (8.47), can be regarded as a mapping

$$\otimes \underbrace{V \times \cdots \times V}_{p \text{ times}} \times \underbrace{V^* \times \cdots \times V^*}_{q \text{ times}} \to T_q^p(V) \tag{8.58}$$

given by

$$\otimes : (v_1, \cdots, v_p, v^1, \cdots, v^q) = v_1 \otimes \cdots \otimes v_p \otimes v^1 \times \cdots \otimes v^q \tag{8.59}$$

for all $v_1, \ldots, v_p \in V$ and $v^1, \cdots, v^q \in V^*$. It is easy to verify that the mapping $\otimes$ is multilinear in the usual sense, i.e.,

$$\begin{aligned} \otimes(v_1, \cdots, \alpha v + \beta u, \ldots, v^q) &= \alpha \otimes (v_1, \ldots, v, \ldots, v^q) \\ &\quad + \beta \otimes (v_1, \ldots, u, \ldots, v^q) \end{aligned} \tag{8.60}$$

where $\alpha v + \beta u$, v and u all take the same position in the argument of $\otimes$ but that position is arbitrary. From (8.60), we see that

$$\begin{aligned} &(\alpha v_1) \otimes v_2 \otimes \cdots \otimes v_p \otimes v^1 \otimes \cdots \otimes v^q \\ &= v_1 \otimes (\alpha v_2) \otimes \cdots \otimes v_p \otimes v^1 \otimes \cdots \otimes v^q \end{aligned}$$

We can extend the operation of tensor product from vectors and covectors to tensors in general. If $A \in T_q^p(V)$ and $B \in T_s^r(V)$, then we define their tensor product $A \otimes B$ to be a tensor of order $(p+r, q+s)$ by

$$\begin{aligned} &A \otimes B(v^1, \ldots, v^{p+r}, v_1, \ldots v_{q+s}) \\ &= A(v^1, \ldots, v^p, v_1, \ldots v_q) B(v^{p+1}, \ldots, v^{p+r}, v_{q+1}, \ldots v_{q+s}) \end{aligned} \tag{8.61}$$

for all $v^1, \ldots, v^{p+r} \in V^*$ and and $v_1, \ldots, v_{q+s} \in V$. Applying this definition to arbitrary tensors A and B yields a mapping

$$\otimes : T_q^p(V) \times T_s^r(V) \to T_{q+s}^{p+r}(V) \tag{8.62}$$

Clearly this operation can be further extended to more than two tensor spaces, say

$$\otimes : T_{q_1}^{p_1}(V) \times \cdots \times T_{q_k}^{p_k}(V) \to T_{q_1 + \cdots + q_k}^{p_1 + \cdots + p_k}(V) \tag{8.63}$$

in such a way that

$$\otimes(A_1, \ldots, A_k) = A_1 \otimes \cdots \otimes A_k \tag{8.64}$$

where $A_i \in T^{p_i}_{q_i}(V)$, i=1,...,k. It is easy to verify that this tensor product operation is also multilinear in the sense generalising (8.60) to tensors. In component form

$$(A \otimes B)^{i_1 \dots i_p}{}_{j_1 \cdots j_{q+s}} = A^{i_1 \dots i_p}{}_{j_1 \dots j_q} B^{i_{p+1} \dots i_{p+r}}{}_{j_{q+1} \cdots j_{q+s}} \tag{8.65}$$

which can be generalised obviously for (8.64) also. From (8.65), or from (8.61), we see that the tensor product is not commutative, but it is associative and distributive.

Relative to a product basis the component form of a tensor $A \in T^p_q(V)$ is given by (8.57) where the components of A can be obtained by (8.55). If we transfer the basis $\{e_i\}$ to $\{\hat{e}_i\}$ as shown by (8.21) and (8.22), then the components of A as well as the product basis relative to {e must be transformed also. The following theorem gives the transformation laws.

Theorem 8.13: Under the transformation of bases (8.21) and (8.22) the product basis (8.14) for $T^p_q(V)$ transforms according to the rule

$$\begin{aligned} &\hat{e}_{i_1} \otimes \cdots \otimes \hat{e}_{i_p} \otimes \hat{e}^{j_1} \otimes \cdots \otimes \hat{e}^{j_q} \\ &\quad = T^{k_1}_{i_1} \cdots T^{k_p}_{i_p} \hat{T}^{j_1}_{l_1} \cdots \hat{T}^{j_q}_{l_p} e_{k_1} \otimes \cdots \otimes e_{k_p} \otimes e^{l_1} \otimes \cdots \otimes e^{l_q} \end{aligned} \tag{8.66}$$

and the components of any tensor $A \in T^p_q(V)$ transform according to the rule.

$$\hat{A}^{i_1 \dots i_p}{}_{j_1 \dots j_q} = \hat{T}^{i_1}_{k_1} \cdots \hat{T}^{i_p}_{k_p} T^{l_1}_{j_1} \cdots T^{l_q}_{j_q} A^{k_1 \dots k_p}{}_{l_1 \dots l_q} \tag{8.67}$$

The proof of these rules involves no more than the multilinearity of the tensor product $\otimes$ and the tensor A. Many classical treatises on tensors use the transformation rule such as (8.67) to define a tensor. The next theorem connects this alternate definition with the one we used.

Theorem 8.14: Given any two sets of $N^{(p+q)}$ scalars $\left\{A^{i_1 \dots i_p}{}_{j_1 \dots j_q}\right\}$ and $\left\{\hat{A}^{i_1 \dots i_p}{}_{j_1 \dots j_q}\right\}$ related by the transformation rule (8.67), there exists a tensor $A \in T_q^p(V)$ whose components relative to the product bases of $\{e_i\}$ to $\{\hat{e}_i\}$ are $\left\{A^{i_1 \ldots i_p}{}_{j_1 \ldots j_q}\right\}$ and $\left\{\hat{A}^{i_1 \ldots i_p}{}_{j_1 \ldots j_q}\right\}$ provided that the bases are related by (8.21) and (8.22).

This theorem is obvious, since we can define the tensor A by (8.57); then the transformation rule (8.67) shows that the components of A relative to the product basis of $\{e_i\}$ are $\left\{A^{i_1 \ldots i_p}{}_{j_1 \ldots j_q}\right\}$. Thus a tensor A corresponds to an equivalence set of components, with the transformation rule serving as the equivalence relation.

As an illustration of the preceding theorem, let us examine whether or not there exists a tensor whose components relative to the product basis of any basis are the values of the generalised Kronecker delta. The answer turns out to be yes, since we have the identify

$$\delta^{i_1 \ldots i_r}_{j_1 \ldots j_r} \equiv \hat{T}^{i_1}_{k_1} \cdots \hat{T}^{i_r}_{k_r} \hat{T}^{l_1}_{j_1} \cdots T^{l_r}_{j_r} \delta^{k_1 \ldots k_r}_{l_1 \ldots l_r} \tag{8.68}$$

which follows from the fact that $\left[T^i_j\right]$ and $\left[\hat{T}^i_j\right]$ are the inverse of each other. We leave the proof of this identity as an exercise for the reader. From Theorem 8.14 and the identity (8.68), we see that there exist a tensor K, of order (r, r) such that

$$K_r = \frac{1}{r!} \delta^{i_1 \ldots i_r}_{j_1 \ldots j_r} e_{i_1} \otimes \cdots \otimes e_{i_r} \otimes e^{j_i} \otimes \cdots \otimes e^{j_r} \tag{8.69}$$

relative to any basis $\{e_i\}$.

By the same line of reasoning, we may ask whether or not there exists a tensor whose components relative to the product basis of any basis are the values of the ε-symbols. The answer turns out to be no, since we have the identities

$$\varepsilon_{i_1 \ldots i_N} = \det\left[\hat{T}^i_j\right] T^{j_l}_{i_l} \cdots T^{j_N}_{i_N} \varepsilon_{j_1 \ldots j_N} \tag{8.70}$$

and

$$\varepsilon^{i_1 \ldots i_N} = \det\left[T^l_j\right] \hat{T}^{i_l}_{j_l} \cdots \hat{T}^{i_N}_{j_N} \varepsilon^{j_1 \ldots j_N} \tag{8.71}$$

which also follows from the fact that $\left[T_j^i\right]$ and $\left[\hat{T}_j^i\right]$ are the inverses of each other.

Since these identities do not agree with the transformation rule (8.67), we can conclude that there exists no tensor whose components relative to the product basis of any basis are always the values of the ε-symbols. In other words, if the values of the ε-symbols are the components of a tensor relative to the product basis of one particular basis $\{e_i\}$, then the components of the same tensor relative to the product basis of another basis generally are not the values of the ε-symbols, unless the transformation matrices $\left[T_j^i\right]$ and $\left[\hat{T}_j^i\right]$ have unit determinant.

In the classical treatises on tensors, the transformation rules (8.70) and (8.71), or more generally

$$A^{i_1 \dots i_p}{}_{j_1 \dots j_q} = \varepsilon \det\left[T_j^i\right]^w \hat{T}_{k_1}^{i_1} \cdots \hat{T}_{k_p}^{i_p} T_{j_1}^{l_1} \cdots T_{j_1}^{l_q} A^{k_1 \dots k_p}{}_{l_1 \dots l_q} \tag{8.72}$$

are used to define *relative tensors*. The exponent w and the coefficient ε on the right-hand side of (8.72) are called the *weight* and the *parity* of the relative tensor, respectively. A relative tensor is called *polar* if its parity has the value +1 in all transformations, while a relative tensor is called axial if ε is equal to the sign of the determinant of the transformation matrix $\left[T_j^i\right]$. In particular, (8.70) shows that $\left\{\varepsilon_{i_1 \dots i_N}\right\}$ are the components of an axial covariant tensor of order N and weight –1, while (8.71) shows that $\left\{\varepsilon^{i_1 \dots i_N}\right\}$ are the components of an axial contravariant tensor of order N and weight +1.

Exercises

1. Prove Theorem 8.11.
2. Prove equation (8.68).
3. Under the canonical isomorphism of $L\,(V;\,V)$ with $T_1^1(V)$, show that an endomorphism $A : V \to V$ and its dual $A^* : V^* \to V^*$ correspond to the same tensor.

4. Define an isomorphism from $L(L(V; V); L(V; V))$ to $T_2^2(V)$ independent of any basis.
5. If $A \in T_2(V)$, show that the determinant of the component matrix $\left[A_{ij}\right]$ of A defines a polar scalar of weight two, i.e., the determinant obeys the transformation rule

$$\det\left[\hat{A}_{ij}\right] = \left(\det\left[T_l^k\right]\right)^2 \det\left[A_{ij}\right]$$

6. Define isomorphisms from $L(V; V^*)$ to; $T_2(V)$, $L(V^*; V)$ to $T^2(V)$, and from $L(V^*; V^*)$ to $T_1^1(V)$ independent of any basis.
7. Given a relation of the form

$$A_{i_1 \ldots i_p klm} B(k, l.m) = C_{i_1 \ldots i_p}$$

where $\left\{A_{i_1 \ldots i_p klm}\right\}$ and $\left\{C_{i_1 \ldots i_p}\right\}$ are components of tensors with respect to any basis, show that the set $\{B(k, l, m)\}$, likewise, can be regarded as components of a tensor, belonging to $T^3(V)$ in this case. This result is known as the *quotient theorem* in classical treatises on tensors.

8. If $A \in T_{p+q}(V)$, define a tensor $T_{pq}A \in T_{p+q}(V)$ by

$$T_{pq}A(u_1, \ldots, u_q, \ldots, v_1, \ldots, v_p) \equiv A(v_1, \ldots, v_p, u_1, \ldots, u_q) \tag{8.73}$$

for all $u_1, \ldots, u_q, v_1, \ldots, v_p \in V$. Show that the operation

$$T_{pq} : T_{p+q}(V) \to T_{p+q}(V) \tag{8.74}$$

is an automorphism of $T_{p+q}(V)$. We call T_{pq} the *generalised transpose operation*. What is the relation between the components of A and $T_{pq}A$?

The Contractions

In this section we shall consider the operation of contracting a tensor of order (p, q) to obtain a tensor of order $(p-1, q-1)$ where p, q are greater than or equal to one. To define this, important operation, we prove first a useful property of the tensor space $T_q^p(V)$. We defined a tensor of order (p, q) to be a multilinear function

$$A:\underbrace{V^*\times\cdots\times V^*}_{p\text{ times}}\times\underbrace{V\times\cdots\times V}_{q\text{ times}}\to R$$

Clearly, this concept can be generalised to a multilinear transformation from the (p, q) -fold Cartesian product of V and V^* to an arbitrary vector space U, namely

$$Z:\underbrace{V\times\cdots\times V}_{p\text{ times}}\times\underbrace{V^*\times\cdots\times V^*}_{q\text{ times}}\to U \tag{8.75}$$

The condition that Z be a multilinear transformation is similar to that for a multilinear function, namely, Z is linear in each one of its variables while its other variables are held constant, e.g., the tensor product $\otimes$ given by (8.58) is a multilinear transformation. The next theorem shows that, in some sense, any multilinear transformation Z of the form (8.75) can be factored through the tensor product $\otimes$ given by (8.58). This fact is known as the *universal factorisation property* of the tensor product.

Theorem 8.15: If Z is an arbitrary multilinear transformation of the form (8.75), then there exists a unique linear transformation

$$C:T_q^p(V)\to U \tag{8.76}$$

such that

$$Z(v_1,\ldots,v_p,v^1,\ldots,v^q)=C(v_1\otimes\cdots\otimes v_p\otimes v^1\otimes\cdots\otimes v^q) \tag{8.77}$$

for all $v_1,\ldots,v_p\in V$ and $v^1,\ldots,v^q\in V^*$.

Proof: Since the simple tensors form a generating set of $T_q^p(V)$, if the linear transformation C satisfying (8.77) exists, then it must be unique. To prove the existence of C, we choose a basis $\{e_i\}$ for V and define the product basis (8.51) for $T_q^p(V)$ as before. Then we define

$$C\left(e_{i_1}\otimes\cdots\otimes e_{i_p}\otimes e^{j_1}\otimes\cdots\otimes e^{j_q}\right)\equiv Z\left(e_{i_1},\ldots,e_{i_p},e^{j_1},\ldots,e^{j_q}\right) \tag{8.78}$$

for all $i_1,\ldots,i_p, j_1,\ldots,j_q=1,\ldots,N$ and extend C to all tensors in $T_q^p(V)$ by linearity. Now it is clear that the linear transformation C defined in this way satisfies the condition (8.77), since both C and Z are multilinear in the vectors $v_1, \ldots v_p$ and the covectors $v^1, \ldots, v^q$, and they agree on the dual bases $\{e_i\}$ and $\{e^i\}$ for V and V^*, as shown in (8.78). Hence they agree on all $(v_1,\ldots, v_p, v^1,\ldots,v^q)$.

If we use the symbol $\otimes$ to denote the multilinear transformation (8.58), then the condition (8.77) can be rewritten as

$$Z = C \circ \otimes \tag{8.79}$$

where the operation $\circ$ on the right-hand side of (8.79) denotes the composition. Equation (8.79) expresses the meaning of the universal factorisation property. In the modern treatises on tensors, this property is often used to define the tensor product and the tensor spaces. Our approach to the concept of a tensor is a compromise between this abstract modern concept and the classical concept based on transformation rules; the preceding theorem and Theorems 8.13 and 8.41 connect our concept with the other two.

Having proved the universal factorisation property of the tensor product, we can now define the operation of contraction. Recall that if $v \in V$ and $v^* \in V^*$, then the scalar product $\langle v, v^* \rangle$ is a scalar. Here we have used the canonical isomorphism given by (8.35). Of course, the, operation

$$\langle , \rangle : V \times V^* \to R$$

is a bilinear function. By Theorem 8.15 $\langle , \rangle$ can be factored through the tensor product

$$\otimes : V \times V^* \to T_1^1(V)$$

i.e., there exists a linear map

$$C : T_1^1(V) \to R \tag{8.80}$$

such that

$$\langle , \rangle = C \circ \otimes$$

or, equivalently,

$$\langle v, v^* \rangle = C(v \otimes v^*) \tag{8.81}$$

for all $v \in V$ and $v^* \in V^*$. This linear function C is the simplest kind of contraction operation. It transforms the tensor space $T_1^1(V)$ to the tensor space $T_{1-1}^{1-1}(V) = T_0^0(V) = R$.

In general if $A \in T_q^p(V)$ is an arbitrary simple tensor, say

$$A = v_1 \otimes \cdots \otimes v_p \otimes v^1 \otimes \cdots \otimes v^p \tag{8.82}$$

then for each pair of integers (i, j), where $1 \le i \le p, 1 \le j \le q$, we seek a unique liner transformation

$$C_j^i : T_q^p(V) \to T_{q-1}^{p-1}(V) \tag{8.83}$$

such that

$$C_j^i A \langle v^j, v_i \rangle v_1 \otimes \cdots \otimes v_{i-1} \otimes v_{i+1} \otimes \cdots \otimes v_p \otimes v^1 \otimes \cdots \otimes v^{j-1} \otimes v^{j+1} \otimes \cdots \otimes v^q \tag{8.84}$$

for all simple tensors A. A more compact notation for the tensor product on the right-hand side is

$$v_1 \otimes \cdots \otimes \breve{v}_i \cdots \otimes v_p \otimes \cdots \otimes v^1 \otimes \breve{v}^j \cdots \otimes v^q \tag{8.85}$$

Since the representation of a simple tensor by a tensor product is not unique, the existence of such a linear transformation C_j^i is by no means obvious. However, we can prove that C_j^i does exist and is uniquely determined by the condition (8.84). Indeed, using the universal factorisation property, we can prove the following theorem.

Theorem 8.16: A unique contraction operation C_j^i satisfying the condition (8.84) exists.

Proof: We define a multilinear transformation of the form (8.75) with $U = T_{q-1}^{p-1}(V)$ by

$$\begin{aligned} &Z(v_1, \ldots, v_p, v^1, \ldots, v^q) \\ &= \langle v^j, v_i \rangle v_1 \otimes \cdots \breve{v}_i \cdots \otimes v_p \otimes v^1 \otimes \cdots \breve{v}^j \cdots \otimes v^q \end{aligned} \tag{8.86}$$

for all $v_1, \ldots, v_p \in V$ and $v^1, v^q \in V^*$. Then by Theorem 8.15 there exists a unique linear transformation C_j^i of the form (8.83) such that (8.84) holds, and thus the proof is complete.

Next we express the contraction operation in component form.

Theorem 8.17: If $A \in T_q^p(V)$ is an arbitrary tensor of order (p, q) then in component form relative to any dual bases $\{e_i\}$ and $\{e^i\}$ we have

$$\left(C_j^i A\right)^{k_1 \ldots \bar{k}_l \ldots k_p}{}_{l_1 \ldots \bar{l}_j \ldots l_q} = A^{k_1 \ldots k_{l-1} t k_{i+1} \ldots k_p}{}_{l_1 \ldots l_{j-1} t l_{j+1} \ldots l_q} \tag{8.87}$$

Proof: Since the contraction operation is linear applying it to the component form (8.57), we get

$$C_j^i A = A^{k_1 \ldots k_p}{}_{l_1 \ldots l_q} \left\langle e^{l_j}, e_{k_i} \right\rangle e_{k_1} \otimes \cdots \breve{e}_{k_i} \cdots \otimes e_{k_p} \otimes e^{l_1} \otimes \cdots \breve{e}^{l_j} \cdots \otimes e^{l_q} \tag{8.88}$$

which means nothing but the formula (8.87) because we have $\left\langle e^{l_j}, e_{k_i} \right\rangle = \delta_{k_i}^{l_j}$.

In particular, for the special case C given by (8.80) we have

$$C(A) = A_t^t \tag{8.89}$$

for all $A \in T_1^1(V)$. If we now make use of the canonical isomorphism

$$T_1^1(V) \cong L(V;V)$$

we see that C coincides with the trace operation defined by (5.70).

Using the contraction operation, we can define a scalar product for tensors. If $A \in T_q^p(V)$ and $B \in T_q^p(V)$, the tensor product $A \otimes B$ is a tensor in $T_{p+q}^{p+q}(V)$ and is defined by (8.61). We apply the contraction

$$C = \underbrace{C_1^1 \circ \cdots \circ C_1^1}_{q\,\text{times}} \circ \underbrace{C_{q+1}^1 \circ \cdots \circ C_{q+1}^1}_{p\,\text{times}} \tag{8.90}$$

to $A \otimes B$, then the result is a scalar $\langle A, B \rangle$, namely

$$\langle A, B \rangle \equiv C(A \otimes B) \tag{8.91}$$

called the scalar product A and B. It is a simple matter to see that $\langle , \rangle$ is a bilinear and definite function in the sense similar to those properties of the scalar product of v and v^*. We can use the bilinear function

$$\langle , \rangle : T_q^p(V) \times T_p^q(V) \to R \tag{8.92}$$

to identify the space $T_q^p(V)$ with the dual space $T_q^p(V)^*$ of the space $T_q^p(V)$ or equivalently, we can define the dual space $T_q^p(V)^*$ abstractly as usual and then introduce a canonical isomorphism from $T_q^p(V)$ to $T_q^p(V)^*$ through (8.91). Thus we write

$$T_q^p(V) \cong T_p^q(V)^* \tag{8.93}$$

Of course we shall also identify the second dual space $T_q^p(V)^{**}$ with $T_q^p(V)$. Hence, we have

$$T_q^p(V)^* \cong T_p^q(V)^{**} \tag{8.94}$$

which follows also by interchanging p and q in (8.93). From (8.81) and (8.90), the scalar product $\langle A, B \rangle$ is given by the component form

$$\langle A, B \rangle = A^{i_1 \dots i_p}{}_{j_1 \dots j_q} B^{j_1 \dots j_p}{}_{i_1 \dots i_p} \tag{8.95}$$

relative to any dual bases $\{e_i\}$ and $\{e^i\}$ for V and V^*

Exercises

1. Show that

   $$\left\langle A, v_1 \otimes \cdots \otimes v_q \otimes v^1 \otimes \cdots \otimes v^p \right\rangle = A\left(v^1, \dots, v^p, v_1, \dots, v_q\right)$$

   for all $A \in T_q^p(V), v_1, \dots, v_q \in V$ and $v^1, \dots, v^p \in V^*$.

2. Give another proof of the quotient theorem by showing that the operation

   $$J : T_s^r(V) \to L\left(T_q^p(V); T_{q-r}^{p-s}(V)\right)$$

defined by

$$(JA)B \equiv \underbrace{C_1^1 \circ \cdots \circ C_1^1}_{s\,\text{times}} \circ \underbrace{C_{s+1}^1 \circ \cdots \circ C_{s+1}^1}_{r\,\text{times}} (A \otimes B)$$

for all $A \in T_s^r(V)$ and $B \in T_q^p(V)$ is an isomorphism. Since there are many such isomorphisms by choosing contractions of pairs of indices differently, we do not make any one of them a canonical isomorphism in general unless stated explicitly.

3. Use the universal factorisation property and prove the existence and the uniqueness of the generalised transpose operation T_{pq} defined by (8.73). *Hint:* Require T_{pq} to be an automorphism of $T_{p+q}(V)$ such that

$$T_{pq}\left(v^1 \otimes \cdots \otimes v^p \otimes u^1 \otimes \cdots \otimes u^q\right) = u^1 \otimes \cdots \otimes u^q \otimes v^1 \otimes \cdots \otimes v^p \tag{8.96}$$

for all simple tensors $v^1 \otimes \cdots \otimes v^p \otimes u^1 \otimes \cdots \otimes u^q \in T_{p+q}(V)$.

4. $\left\{e_{i_1} \otimes \cdots \otimes e_{i_p} \otimes e^{j_1} \otimes \cdots \otimes e^{j_q}\right\}$ is the product basis for $T_p^q(V)$ induced by the dual bases $\{e_i\}$ and $\{e^i\}$; construct its dual basis in $T_p^q(V)$ with respect to the scalar product defined by (8.91).

Tensors on Inner Product Spaces

The main result of this section is that if V is equipped with a particular inner product, then we can identify the tensor spaces of the same total order by means of various canonical isomorphisms, a special case of these being the isomorphism G from V to V^*.

Recall that the isomorphism $G : V \to V^*$ is defined by the condition

$$\langle Gu, v\rangle = \langle Gv, u\rangle = (u.v), \quad u, v \in V \tag{8.97}$$

In general, if A is a simple, pure, contravariant tensor of order p, say

$$A = v_1 \otimes \cdots \otimes v_p$$

then we define the *pure covariant representation* of A to be the tensor

$$G^p A = Gv_1 \otimes \cdots \otimes Gv_p \in T_p(V) \tag{8.98}$$

By linearity, G^p can be extended to all of $T^p(V)$. Equation (8.98) means that

$$G^p A\left(u_1, \cdots, u_p\right) = \langle Gv_1, u\rangle \cdots \left\langle Gv_p, u_p\right\rangle = \left(v_1 . u_1\right) \cdots \left(v_p . u_p\right) \tag{8.99}$$

for all $u_1, \ldots, u_p \in V$. Equation (8.99) generalises the condition (8.97) from v to $v_1 \otimes \cdots \otimes v_p$. Of course, because the tensor product is not one-to-one, we have to show that the covariant tensor $G^p A$ does not depend on the representation of A. This fact follows directly from the universal factorisation of the tensor product. Indeed, if we define a p-linear map

$$Z : \underbrace{V \times \cdots \times V}_{p\,\text{times}} \to T_p(V)$$

by

$$Z(v_1, \ldots, v_p) \equiv Gv_1 \otimes \cdots \otimes Gv_p \tag{8.100}$$

then Z can be factored through the tensor product $\otimes$, i.e., there exists a unique linear transformation G^p from $T^p(V)$ to $T_p(V)$,

$$G^p : T^p(V) \to T_p(V) \tag{8.101}$$

such that

$$Z = G^p \circ \otimes$$

or, equivalently,

$$Z(v_1, \ldots, v_p) \equiv G^p(v_1 \otimes \cdots \otimes v_p) G^p A \tag{8.102}$$

Comparing (8.102) with (8.100) we see that G^p obeys the condition (8.98), and thus $G^p A$ is well defined by (8.99).

It is easy to verify that G^p is an isomorphism. Indeed $(G^p)^{-1}$ is the unique liner transformation from $T_p(V)$ to $T^p(V)$ such that

$$(G^p)^{-1}\left(v^1 \otimes \cdots \otimes v^p\right) = G^{-1}v^1 \otimes \cdots \otimes G^{-1}v^p \tag{8.103}$$

for all $v^1, \ldots, v^p \in V^*$. Thus G^p makes $T^p(V)$ and $T_p(V)$ isomorphic, just as G makes V and V^* isomorphic. In fact, $G = G^1$.

Clearly, we can extend the preceding argument to mixed tensor spaces on V also. If A is a mixed simple tensor in $T_q^p(V)$, say

$$A = v_1 \otimes \cdots \otimes v_p \otimes u^1 \otimes \cdots \otimes u^q \tag{8.104}$$

then we define the *pure covariant representation* of A to be the tensor

$$G_q^p A = Gv_1 \otimes \cdots \otimes Gv_p \otimes u_1 \cdots \otimes u^q \tag{8.105}$$

and G_q^p is an isomorphism from $T_q^p(V)$ to $T_{p+q}(V)$,

$$G_q^p : T_q^p(V) \to T_{p+q}(V) \tag{8.106}$$

Indeed, its inverse is characterised by

$$\begin{aligned} &\left(G_q^p\right)^{-1}(v^1 \otimes \cdots \otimes v^p \otimes u^1 \otimes \cdots \otimes u^q) \\ &= G^{-1}v^1 \otimes \cdots \otimes G^{-1}v^p \otimes u^1 \otimes \cdots \otimes u^q \end{aligned} \tag{8.107}$$

for all $v^1, \ldots, v^p, u^1, \ldots, u^q \in V^*$. Clearly, by suitable compositions of the operations G_q^p and their inverses, we can define isomorphisms between tensor spaces of the same total order. For example, if p_1 and q_1 are another pair of integers such that

$$p_1 + q_1 = p + q$$

then $T_q^p(V)$ is isomorphic to $T_{q_1}^{p_1}(V)$ by the isomorphism

$$\left(G_{q_1}^{p_1}\right)^{-1} \circ G_q^p : T_q^p(V) \to T_{q_1}^{p_1}(V) \tag{8.108}$$

In particular, if $q_1 = 0$, $p_1 = p + q$, then for any $A \in T_q^p(V)$ the tensor $\left(G^{p+q}\right)^{-1} G_q^p A \in T^{p+q}(V)$ is called the *pure contravariant* representation of A. For example, if A is given by (8.104), then its pure contravariant representation is given by

$$\left(G^{p+q}\right)^{-1} G_q^p A = v_1 \otimes \cdots \otimes v_p \otimes G^{-1}u^{-1} \otimes \cdots \otimes G^{-1}u^q$$

We shall now express the isomorphism G_q^p in component form. If $A \in T_q^p(V)$ has the component representation

$$A = A^{i_1 \dots i_p}{}_{j_1 \dots j_q} e_{i_1} \otimes \cdots \otimes e_{i_p} \otimes e^{j_1} \otimes \cdots \otimes e^{j_q} \tag{8.109}$$

then from (8.105) we obtain

$$G_q^p A = A^{i_1 \dots i_p}{}_{j_1 \dots j_q} Ge_{i_1} \otimes \cdots \otimes Ge_{i_p} \otimes e^{j_1} \otimes \cdots \otimes e^{j_q} \tag{8.110}$$

This result can be written as

$$G_q^p A = A^{i_1 \dots i_p}{}_{j_1 \dots j_q} \overline{e}_{i_1} \otimes \cdots \otimes \overline{e}_{i_p} \otimes e^{j_1} \otimes \cdots \otimes e^{j_q} \tag{8.111}$$

where $\{\overline{e}_i\}$ is a basis in V^* reciprocal to $\{e^i\}$, i.e.,

$$\overline{e}_i . e^j = \delta_i^j \tag{8.112}$$

since it follows from (5.63), (4.14), and (8.97) that we have

$$\overline{e}_i = Ge_i = e_{ji} e^j \tag{8.113}$$

where

$$e_{ji} = e_j . e_i \tag{8.114}$$

Substituting (8.113) into (8.111), we obtain

$$G_q^p A = A_{i_1 \dots i_p j_1 \dots j_q} e^{i_1} \otimes \cdots \otimes e^{i_p} \otimes e^{j_1} \otimes \cdots \otimes e^{j_q} \tag{8.115}$$

where

$$A_{i_1 \dots i_p j_1 \dots j_q} = e_{i_1 k_1} \cdots e_{i_p k_p} A^{k_1 \dots k_p}{}_{j_1 \dots j_q} \tag{8.116}$$

This equation illustrates the component form of the isomorphism G_q^p. It has the effect of *lowering* the first p superscripts on the components $\left\{A^{i_1 \dots i_p}{}_{j_1 \dots j_q}\right\}$. Thus $A \in T_q^p(V)$ and $G_q^p A \in T_{p+q}(V)$ have the same components if the bases in (8.109) and (8.111) are used. On the other hand, if the usual product basis for $T_q^p(V)$ and $T_{p+q}(V)$ are used, as in (8.109) and (8.115), the component of A and $G_q^p A$ are related by (8.116).

Since G_q^p is an isomorphism, its inverse exists and is a linear transformation from $T_{p+q}(V)$ and $T_q^p(V)$. If $A \in T_{p+q}(V)$ has the component representation

$$A = A_{i_1 \dots i_p j_1 \dots j_q} \otimes \cdots \otimes e^{i_p} \otimes e^{j_1} \otimes \cdots \otimes e^{j_q} \tag{8.117}$$

then from (8.107)

$$\left(G_q^p\right)^{-1} A = A_{i_1 \dots i_p j_1 \dots j_q} G^{-1} e^{i_1} \otimes \cdots \otimes G^{-1} e^{i_p} \otimes e^{j_1} \otimes \cdots \otimes e^{j_q} \tag{8.118}$$

By the same argument as before, this formula can be rewritten as

$$\left(G_q^p\right)^{-1} A = A_{i_1 \dots i_p j_1 \dots j_q} \overline{e}^{i_1} \otimes \cdots \otimes \overline{e}^{i_p} \otimes e^{j_1} \otimes \cdots \otimes e^{j_q} \tag{8.119}$$

where $\left\{\overline{e}^i\right\}$ is a basis of V reciprocal to $\{e_i\}$, or equivalently as

$$\left(G_q^p\right)^{-1} A = A^{i_1 \dots i_p}{}_{j_1 \dots j_q} e_{i_1} \otimes \cdots \otimes e_{i_p} \otimes e^{j_1} \otimes \cdots \otimes e^{j_q} \tag{8.120}$$

where

$$A^{i_1 \dots i_p}{}_{j_1 \dots j_q} = e^{i_1 k_1} \cdots e^{i_p k_p} A_{k_1 \dots k_p j_1 \dots j_q} \tag{8.121}$$

Here of course $\left[e^{ij}\right]$ is the inverse matrix of $\left[e_{ij}\right]$ and is given by

$$e^{ij} = e^i . e^j \tag{8.122}$$

since

$$\overline{e}^i = G^{-1} e^i = e^{ij} e_j \tag{8.123}$$

Equation (8.121) illustrates the component form of the isomorphism $\left(G_q^p\right)^{-1}$. It has the effect of raising the first p subscripts on the components $\left\{A_{i_1 \dots i_p j_1 \dots j_q}\right\}$.

Combining (8.116) and (8.121), we see that the component form of the isomorphism $\left(G_{q_1}^{p_1}\right)^{-1} \circ G_q^p$ in (8.108) is given by raising the first p_1–p superscripts of $A \in T_q^p(V)$ if $p_1 > p$, or by lowering the last p–p_1 subscripts of A if $p > p_1$. Thus if A has the component form (8.109), then $\left(G_{q_1}^{p_1}\right)^{-1} G_q^p A$ has the component form

$$\left(G_{q_1}^{p_1}\right)^{-1} G_q^p A = A^{i_1 \dots i_{p_1}}{}_{j_1 \cdots j_{q_1}} e_{i_1} \otimes \cdots \otimes e_{i_{p_1}} \otimes e^{j_1} \otimes \cdots \otimes e^{j_{q_1}} \quad (8.124)$$

where

$$A^{i_1 \dots i_{p_1}}{}_{j_1 \cdots j_{q_1}} = A^{i_1 \dots i_p}{}_{k_1 \dots kp_1 - p^{j_1} \dots jq_1} e^{i_{p+1}k_1} \cdots e^{i_q k_{p_1 - p}} \text{ if } p_1 > p \quad (8.125)$$

and

$$A^{i_1 \dots i_{p_1}}{}_{j_1 \cdots j_{q_1}} = A^{i_1 \dots i_{p_1} \dots k_{p-p_1}}{}_{j_{p-p_1+1} \cdots j_{q_1}} e_{k_1 j_1} \cdots e_{k_{p-p_1} j_{p-p_1}} \text{ if } p > p_1 \quad (8.126)$$

For example, if $A \in T_2^1(V)$ has the representation

$$A = A^i{}_{jk} e_i \otimes e^j \otimes e^k$$

then the covariant representation of A is

$$G_2^1 A = A^i{}_{jk} \overline{e}_i \otimes e^j \otimes e^k = A^i{}_{jk} e_{il} e^l \otimes e^j \otimes e^k = A_{ljk} e^l \otimes e^j \otimes e^k$$

the contravariant representation of A is

$$\left(G^3\right)^{-1} G_2^1 A = A^i{}_{jk} e_i \otimes \overline{e}^j \otimes \overline{e}^k = A^i{}_{jk} e^{jl} e^{km} e_i \otimes e_l \otimes e_m \equiv A^{ilm} e_i \otimes e_l \otimes e_m$$

and the representation of A in $T_1^2(V)$ is

$$\left(G_1^2\right)^{-1} G_2^1 A = A^i{}_{jk} e_i \otimes \overline{e}^j \otimes e^k = A^i{}_{jk} e^{jl} e_i \otimes e_l \otimes e^k \equiv A^{il}{}_k e_i \otimes e_l \otimes e^k$$

etc. These formulas follow from (8.116), (8.120), and (8.125).

Of course, we can apply the operations of lowering and raising to indices at any position, e.g., we can define a "component" $A^{i_1}{}_{j_2}{}^{i_3 \dots i_a \dots}{}_{\dots j_b \dots}$ for $A \in T_r(V)$ by

$$A^{i_1}{}_{j_2}{}^{i_3 \dots i_a \dots}{}_{\dots j_b \dots} \equiv A_{j_1 j_2 \dots j_r} e^{i_1 j_1} e^{i_3 j_3} \cdots e^{i_a j_a} \cdots \quad (8.127)$$

However, for simplicity, we have not yet assigned any symbol to such representations whose "components" have an irregular arrangement of superscripts and subscripts. In our notation for the component representation of a tensor $A \in T_q^p(V)$ the contravariant superscripts always come first, so that the components of A are written as $A^{i_1 \dots i_p}{}_{j_1 \dots j_q}$ as shown in (8.109), not as $A_{j_1 \dots j_q}{}^{i_1 \dots i_p}$ or as any other rearrangement of positions of the superscripts $i_1 \ldots i_p$ and the subscripts $j_1 \ldots j_q$, such as the one defined by (8.127). In order to indicate precisely the position of the contravariant indices and the covariant indices in the irregular component form, we may use, for example, the notation

$$\underset{(1)}{V} \otimes \underset{(2)}{V^*} \otimes \underset{(3)}{V} \otimes \cdots \otimes \underset{(a)}{V} \otimes \cdots \otimes \underset{(b)}{V^*} \otimes \cdots \tag{8.128}$$

for the tensor space whose elements have components of the form on the left-hand side of (8.127), where the order of V and V^* in (8.128) are the same as those of the contravariant and the covariant indices, respectively, in the component form. In particular, the simple notation $T_q^p(V)$ now corresponds to

$$\underset{(1)}{V} \otimes \cdots \otimes \underset{(p)}{V} \otimes \underset{(p+1)}{V^*} \otimes \cdots \otimes \underset{(p+q)}{V^*} \tag{8.129}$$

Since the irregular tensor spaces, such as the one in (8.128), are not convenient to use, we shall avoid them as much as possible, and we shall not bother to generalise the notion G_q^p to isomorphisms from the irregular tensor spaces to their corresponding pure covariant representations.

So far, we have generalised the isomorphism G for vectors to the isomorphisms G_q^p for tensors of type (p, q) in general. Equation (8.97) for G, however, contains more information than just the fact that G is an isomorphism. If we read that equation reversely, we see that G can be used to compute the inner product on V. Indeed, we can rewrite that equation as

$$v.u = \langle Gv, u \rangle \equiv \left\langle G^1 v, u \right\rangle \tag{8.130}$$

This idea can be generalised easily to tensors. For example, if A and B are tensors in $T^r(V)$, then we can define an inner product $A.B$ by

$$A.B \equiv \left\langle G^r A, B \right\rangle \tag{8.131}$$

We leave the proof to the reader that (8.131) actually defines an inner product $T^r(V)$. By use of (8.95) and (8.116), it is possible to write (8.131) in the component form

$$A.B = A_{j_1 \ldots j_r} e^{i_1 j_1} \cdots e^{i_r j_r} B_{i_1 \ldots i_r} \tag{8.132}$$

or, equivalently, in the forms

$$A.B = \begin{cases} A^{i_1 \ldots j_1} B_{i_1 \ldots i_r} \\ A^{i_1 j_{r-1}}{}_{j_r} e^{i_r j_r} B_{i_1 \ldots i_r} \\ A^{i_1 \ldots i_r} B^{j_1}{}_{i_2 \ldots i_r} e_{i_1 j_1}, \qquad \text{etc.} \end{cases} \tag{8.133}$$

Equation (8.133) suggests definitions of inner products for other tensor spaces besides $T^r(V)$. If A and B are in $T_q^p(V)$, we define $A.B$ by

$$A.B \equiv \left(G^{p+q}\right)^{-1} G_q^p A.\left(G^{p+q}\right)^{-1} G_q^p B = \left\langle G_q^p A, \left(G^{p+q}\right)^{-1} G_q^p B \right\rangle \tag{8.134}$$

Again, we leave it as an exercise to the reader to establish that (8.134) does define an inner product on $T_q^p(V)$; moreover, the inner product can be written in component form by (8.133) (8.133) also.

We have pointed out that if we agree to use a particular inner product, then the isomorphism G can be regarded as canonical. The formulas of this section clearly indicate the desirability of this procedure if for no other reason than notational simplicity. Thus, from this point on, we shall identify the dual space V^* with V, i.e.,

$$V \cong V^*$$

by suppressing the symbol G. Then in view of (8.98) and (8.103), we shall suppress the symbol G_q^p also. Thus, we shall identify all tensor spaces of the same total order, i.e., we write

$$T^r(V) \cong T_1^{r-1}(V) \cong \cdots \cong T_r(V) \tag{8.135}$$

By this procedure we can replace scalar products throughout our formulas by inner products according to the formulas (8.9) and (8.134).

The identification (8.135) means that, as long as the total order of a tensor is given, it is no longer necessary to specify separately the contravariant order and the covariant order. These separate orders will arise only when we select a particular component representation of the tensor. For example, if A is of total order r, then we can express A by the following different component forms:

$$\begin{aligned} A &= A^{i_1 \ldots i_r} e_{i_1} \otimes \cdots \otimes e_{i_r} \\ &= A^{i_1 \ldots i_{r-1}}{}_{j_r} e_{i_1} \otimes \cdots \otimes e_{i_{r-1}} \otimes e^{j_r} \\ &\vdots \\ &= A_{j_1 \ldots j_r} e^{j_1} \otimes \cdots \otimes e^{j_r} \end{aligned} \tag{8.136}$$

where the placement of the indices indicates that the first form is the pure contravariant representation, the last form is the pure covariant representation, while the intermediate forms are various mixed tensor representations, all having the same total order r. Of course, the various representations in (8.136) are related by the formulas (8.116), (8.121), (8.125), and (8.126). In fact, if (8.127) is used, we can even represent A by an irregular component form such as

$$A = A^{i_1 i_3 \ldots i_a}{}_{j_2 \ldots j_b \ldots} e_{i_1} \otimes e^{j_2} \otimes e_{i_3} \otimes \cdots \otimes e_{i_a} \otimes \cdots \otimes e^{j_b} \otimes \cdots \tag{8.137}$$

provided that the total order is unchanged. The contraction operator can now be rewritten in a form more convenient for tensors defined on an inner product space. If A is a simple tensor of (total) order $r, r \geq 2$, with the representation

$$A = v_1 \otimes \cdots \otimes v_r$$

then $C_{ij}A$, where $1 \leq i < j \leq r$, is a simple tensor of (total) order r-2 defined by (8.84),

$$C_{ij}A \equiv (v_i . v_j) v_1 \otimes \cdots \breve{v}_1 \cdots \breve{v}_j \cdots \otimes v_r \tag{8.138}$$

By linearity, C_{ij} can be extended to all tensors of order r. If $A \in T_r(V)$ has the representation

$$A = A_{k_1 \ldots k_r} e^{k_1} \otimes \cdots \otimes e^{k_r}$$

then by (8.138) and (8.122)

$$C_{ij}A = A_{k_1 \ldots k_r} C_{ij}(e^{k_1} \otimes \cdots \otimes e^{k_r})$$
$$= e^{k_i k_j} A_{k_1 \ldots} e^{k_1} \otimes \cdots \breve{e}^{k_i} \cdots \breve{e}^{k_j} \cdots \otimes e^{k_r}$$
$$= A_{k_1 \ldots \underset{i}{\ldots} \underset{j}{k_j} \ldots k_r}^{\quad\; k_j} e^{k_1} \otimes \cdots \breve{e}_{k_i} \cdots \breve{e}^{k_j} \cdots \otimes e^{k_r} \tag{8.139}$$

It follows from the definition (8.138) that the complete contraction operator C

$$C : T_{2r}(V) \to R$$

can be written

$$C = C_{12} \circ C_{13} \circ \cdots \circ C_{1(r+1)} \tag{8.140}$$

Also, if A and B are in $T_r(V)$, it is easily establish that

$$A.B = C(A \otimes B) \tag{8.141}$$

In closing this chapter, it is convenient for later use to record certain formulas here. The identity automorphism I of V corresponds to a tensor of order 2. Its pure covariant representation is simply the inner product

$$I\ (u,\ v) = u.v \tag{8.142}$$

The tensor I can be represented in any of the following forms:

$$I = e_{ij} e^i \otimes e^j = e^{ij} e_i \otimes e_j = \delta^i_j e_i \otimes e^j = \delta^j_i e^i \otimes e_j \tag{8.143}$$

There is but one contraction for I, namely

$$C_{12} I = trI = \delta^i_i = N \tag{8.144}$$

In view of (8.142), the identity tensor is also called the *metric tensor* of the inner product space.

Exercises

1. Show that (8.141) defines an inner product on T_r(V)
2. Show that the formula (8.134) can be rewritten as

$$A.B = \left\langle (G_p^q)^{-1} T_{pq} G_q^p A, B \right\rangle \tag{8.145}$$

where T_{pq} is the generalised transpose operator. In particular, if A and B are second order tensors, then (8.145) reduces to the more familiar formula:

$$A.B = tr(A^T B) \tag{8.146}$$

3. Show that the linear transformation

   $$G^p : T^p(V) \to T_p(T)$$

   defined by (8.98) can also be characterised by

   $$(G^p A)(u_1, \ldots, u_p) \equiv A(Gu_1, \ldots, Gu_p) \tag{8.147}$$

   for all $A \in T^p(V)$ and all $u_1, \ldots, u_p \in V$.

4. Show that if A is a second-order tensor, then

   $$A = C_{23}(I \otimes A)$$

   What is the component form of this identity?

Exterior Algebra

The purpose of this chapter is to formulate enough machinery to define the determinant of an endomorphism in a component free fashion, to introduce the concept of an orientation for a vector space, and to establish a certain isomorphism which generalises the classical operation of vector product to a N-dimensional vector space. For simplicity, we shall assume throughout this chapter that the vector space, and thus its associated tensor spaces, are equipped with an inner product.

Skew-Symmetric Tensors and Symmetric Tensors

If $A \in T_r(V)$ and σ is a given permutation of $\{1,...,r\}$ then we can define a new tensor $T_\sigma A \in T_r(V)$ by the formula

$$T_\sigma A(v_1,...,v_r) = A(v_{\sigma(1)},...,v_{\sigma(r)}) \tag{9.1}$$

for all $v_1,...,v_r \in V$. For example, the generalised transpose operation T_{pq} defined by (8.73) is a special case of T_σ with σ given by

$$\sigma = \begin{pmatrix} 1 & \cdot & \cdot & \cdot & q & q+1 & \cdot & \cdot & \cdot & q+p \\ p+1 & \cdot & \cdot & \cdot & p+q & 1 & \cdot & \cdot & \cdot & p \end{pmatrix}$$

Naturally, we call $T_\sigma A$ the σ*-transpose* of A for any σ in general. We have obtained the components of the transpose T_{pq}.

For an arbitrary permutation σ the components of $T_\sigma A$ are related to those of A by

$$(T_\sigma A)_{i_1 \dots i_r} = A_{i_{\sigma(1)} \dots i_{\sigma(r)}} \tag{9.2}$$

Using the same argument, we can characterise T_σ by the condition that

$$T_\sigma(v^1 \otimes \cdots \otimes v^r) = v^{\sigma^{-1}(1)} \otimes \cdots \otimes v^{\sigma^{-1}(r)} \tag{9.3}$$

for all simple tensors $v^1 \otimes \cdots \otimes v^r \in T_r(V)$

If $T_\sigma A = A$ for all permutations σA is said to be (*completely*) *symmetric*. On the other hand if $T_\sigma A = \varepsilon_\sigma A$ for all permutations σ, where ε_σ denotes the parity of σ, then A is said to be (*completely*) *skew-symmetric*. For example, the identity tensor I given by (8.126) is a symmetric second-order tensor, while the tensor $u \otimes v - v \otimes u$, for any vectors $u, v \in V$, is clearly a skew-symmetric second-order tensor. We shall denote by $\hat{T}_r(V)$ the set of all skew-symmetric tensors in $T_r(V)$. We leave it to the reader to establish the fact that $\hat{T}_r(V)$ is a subspace of $T_r(V)$. Elements of $\hat{T}_r(V)$ are often called *r-vectors or r-forms.*

Theorem 9.1: An element $A \in \hat{T}_r(V)$ assumes the value 0 if any two of its variables coincide.

Proof: We wish to establish that

$$A\,(v_1, \dots, v, \dots, v, \dots, v_r) = 0 \tag{9.4}$$

This result is a special case of the formula

$$A\,(v_1, \dots, v_s, \dots, v_t, \dots, v_r) = -A\,(v_1, \dots, v_t, \dots, v_s, \dots, v_r) = 0 \tag{9.5}$$

which follows by the fact that $\varepsilon_\sigma = -1$ for the permutation which switches the pair of indices (s, t) while leaving the remaining indices unchanged. If we take $v = v_s = v_t$ in (9.5), then (9.4) follows.

Corollary: An element $A \in \hat{T}_r(V)$ assumes the value zero if it is evaluated on a linearly dependent set of vectors.

This corollary generalises the result of the preceding theorem but is itself also a direct consequence of that theorem, for if $\{v_1,...,v_r\}$ is a linearly dependent set, then at least one of the vectors can be expressed as is a linear combination of the remaining ones. From the r-linearity of A, $A\ (v_1,...,v_r)$ can then be written as the linear combination of quantities which are all equal to zero because which are the values of A at arguments having at least two equal variables.

Corollary: If .r is greater than N, the dimension of V, then $\hat{T}_r(V) = \{0\}$.

This corollary follows from the last corollary and the fact that every set of more than N vectors in a N-dimensional space is linearly dependent.

Exercises

1. Show that the set of symmetric tensors of order r forms a subspace of $T_r(V)$.
2. Show that $T_{\sigma\tau} = T_\sigma T_\tau$ for all permutations σ and τ. Also, show that T_σ is the identity automorphism of $T_r(V)$ if and only if σ is the identity permutation.

Skew-Symmetric Operator

In this section we shall construct a projection from $T_r(V)$ into $T_r(V)$. This projection is called the *skew-symmetric operator*. If $A \in T_r(V)$, we define the *skew-symmetric projection* K_rA of A by

$$K_rA = \frac{1}{r!}\sum_{\sigma} \varepsilon_\sigma T_\sigma A \tag{9.6}$$

where the summation is taken overall permutations σ of $\{1,...,r\}$. The endomorphism

$$K_r : T_r(V) \rightarrow T_r(V) \tag{9.7}$$

defined in this way is called the *skew-symmetric operator*.

Before showing that K_r has the desired properties, we give one example first. For simplicity, let us choose $r = 2$. Then there are only two permutations of $\{1, 2\}$, namely

$$\sigma = \begin{pmatrix} 1 & 2 \\ 1 & 2 \end{pmatrix}, \quad \sigma = \begin{pmatrix} 1 & 2 \\ 2 & 1 \end{pmatrix} \tag{9.8}$$

and their parities are

$$\varepsilon_\sigma = 1 \text{ and } \varepsilon_\sigma = -1 \tag{9.9}$$

Substituting (9.8) and (9.9) into (9.6), we get

$$(K_2A)(v_1, v_2) = \frac{1}{2}\left(A(v_1, v_2) + A(v_2, v_1)\right) \tag{9.10}$$

for all $A \in T_2(V)$ and $v_1, v_2 \in V$. In particular, if A is skew-symmetric, namely

$$A(v_2, v_1) = -A(v_1, v_2) \tag{9.11}$$

then (9.10) reduces to

$$(K_2A)(v_1, v_2) = A(v_1, v_2)$$

or, equivalently

$$K_2A = A, \qquad A \in \hat{T}_2(V) \tag{9.12}$$

Since from (9.10), $K_2A \in \hat{T}_2(V)$ for any $A \in T_2(V)$????,by (9.12) we then have

$$K_2(K_2A) = K_2A$$

or, equivalently,

$$K_2^2 = K_2 \tag{9.13}$$

which means that K_2 is a projection. Hence, for second order tensors, K_2 has the desired properties. We shall now prove the same for tensors in general.

Theorem 9.2: Let $K_r : T_r(V) \to T_r(V)$ be defined by (9.6). Then the range of K_r is $\hat{T}_r(V)$, namely

$$R(K_r) = \hat{T}_r(V) \tag{9.14}$$

Moreover, the restriction of K_r on $\hat{T}_r(V)$ is the identity automorphism of $\hat{T}_r(V)$, i.e.,

$$K_r A = A, \qquad A \in \hat{T}_r(V) \tag{9.15}$$

Proof: We prove the equation (9.15) first. If $A \in \hat{T}_r(V)$, then by definition we have

$$T_\sigma A = \varepsilon_\sigma A \tag{9.16}$$

for all permutations σ. Substituting (9.16) into (9.6), we get

$$K_r A = \frac{1}{r!}\sum_\sigma \varepsilon_\sigma^2 A = \frac{1}{r!}(r!)A = A \tag{9.17}$$

Here we have used the familiar fact that there are a total of $r!$ permutations for r numbers $\{1,\ldots, r\}$.

Having proved (9.15), we can conclude immediately that

$$R(K_r) \supset \hat{T}_r(V) \tag{9.18}$$

since $\hat{T}_r(V)$ is a subspace of $T_r(V)$. Hence, to complete the proof it suffices to show that

$$R(K_r) \subset \hat{T}_r(V)$$

This condition means that

$$T_\tau(K_r A) = \varepsilon_\tau K_r A \tag{9.19}$$

for all $A \in T_r(V)$. From (9.6), $T_\tau(K_r A)$ is given by

$$T_\tau(K_r A) = \frac{1}{r!}\sum_\sigma \varepsilon_\sigma T_\tau(T_\sigma A)$$

We can rewrite the preceding equation as

$$T_\tau(K_r A) = \frac{1}{r!}\sum_\sigma \varepsilon_\sigma T_{\tau\sigma} A \tag{9.20}$$

Since the set of all permutations form a group, and since

$$\varepsilon_\tau \varepsilon_\sigma = \varepsilon_{\tau\sigma}$$

the right-band side of (9.20) is equal to

$$\frac{1}{r!}\sum_\sigma \varepsilon_\tau \varepsilon_{\tau\sigma} T_{\tau\sigma} A$$

or, equivalently,

$$\varepsilon_\tau \frac{1}{r!}\sum_\sigma \varepsilon_{\tau\sigma} T_{\tau\sigma} A$$

which is simply another way of writing $\varepsilon_\tau K_r A$, so (9.19) is proved.

By exactly the same argument leading to (9.19) we can prove also that

$$K_r(T_\tau A) = \varepsilon_\tau K_r A \tag{9.21}$$

Hence K_r and T_τ commute for all permutations τ. Also, (9.14) and (9.15) now imply that $K_r^2 = K_r$, for all r. Hence we have shown that K_r is a projection from $T_r(V)$ to $\hat{T}_r(V)$. From a result for projections in general, $T_r(V)$ can be decomposed into the direct sum

$$T_r(V) = \hat{T}_r(V) \oplus K(K_r) \tag{9.22}$$

where $K(K_r)$ is the: kernel of K_r and is characterised by the following theorem.

Theorem 9.3: The kernel $K(K_r)$ is generated by the set of simple tensors $v^1 \otimes \cdots \otimes v^r$ having at least one pair of equal vectors among the vectors $v^1,\ldots,v^r$.

Proof: Since $T_r(V)$ is generated by simple tensors, it suffices to show that the difference

$$v^1 \otimes \cdots \otimes v^r - K_r(v^1 \otimes \cdots \otimes v^r) \tag{9.23}$$

can be expressed as a linear combination of simple tensors having the prescribed property. From (9.3) and (9.6), the difference (9.23) can be written as

$$\frac{1}{r!}\sum_\sigma (v^1 \otimes \cdots \otimes v^r - \varepsilon_\sigma v^{\sigma(1)} \otimes \cdots \otimes v^{\sigma(r)}) \tag{9.24}$$

We claim that each sum of the form

$$v^1 \otimes \cdots \otimes v^r - \varepsilon_\sigma v^{\sigma(1)} \otimes \cdots \otimes v^{\sigma(r)} \tag{9.25}$$

can be expressed as a sum of simple tensors each having at least two equal vectors among the r vectors forming the tensor product. We have mentioned that every permutation σ can be decomposed into a product of permutations each switching only one pair of indices, say

$$\sigma = \sigma_k \sigma_{k-1} \cdots \sigma_2 \sigma_1 \tag{9.26}$$

where the number k is even or odd corresponding to σ being an even or odd permutation, respectively. Using the decomposition (9.26), we can rewrite (9.25) in the form

$$\begin{aligned}
&+\left(v^1 \otimes \cdots \otimes v^r + v^{\sigma_1(1)} \otimes \cdots \otimes v^{\sigma_1(r)}\right) \\
&-\left(v^{\sigma_1(1)} \otimes \cdots \otimes v^{\sigma_1(r)} + v^{\sigma_2\sigma_1(1)} \otimes \cdots \otimes v^{\sigma_2\sigma_1(r)}\right) \\
&+\cdots \\
&-\cdots \\
&-\varepsilon_\sigma \left(v^{\sigma_{k-1}\ldots\sigma_1(1)} \otimes \cdots \otimes v^{\sigma_{k-1}\ldots\sigma_1(r)} + v^{\sigma(1)} \otimes \cdots \otimes v^{\sigma(r)}\right)
\end{aligned} \tag{9.27}$$

where all the intermediate terms $v^{\sigma_1\ldots\sigma_r(1)} \otimes \cdots \otimes v^{\sigma_l\ldots\sigma_1(r)}, j=1,\ldots,k-1$ cancel in the sum. Since each of $\sigma_1,\ldots,\sigma_k$ switches only one pair of indices, a typical term in the sum (9.27) has the form

$$v^a \otimes \cdots v^s \cdots v^t \cdots \otimes v^b + v^a \otimes \cdots v^t \cdots v^s \cdots \otimes v^b$$

which can be combined into three simple tensors:

$$v^a \otimes \cdots (u) \cdots (u) \cdots \otimes v^b$$

where $u = v^s$, v^t, and $v^s + v^t$. Consequently, (9.27) is a sum of simple tensors having the property prescribed by the theorem. Of course, from (9.21) all those simple tensors belong to the kernel K (K_r). The proof is complete.

In view of the decomposition (9.22), the subspace $\hat{T}_r(V)$ is isomorphic to the factor space $T_r(V)/K(K_r)$. In fact, some authors use this structure to define the space $\hat{T}_r(V)$ abstractly without making $\hat{T}_r(V)$ a subspace of $T_r(V)$. The preceding theorem shows that this abstract definition of $\hat{T}_r(V)$ is equivalent to ours.

The next theorem gives a useful property of the skew-symmetric operator K_r.

Theorem 9.4: If $A \in T_p(V)$ and $B \in T_p(V)$, then

$$\begin{aligned}
K_{p+q}(A \otimes B) &= K_{p+q}(A \otimes K_q B) \\
&= K_{p+q}(K_p A \otimes B) \\
&= K_{p+q}(K_p A \otimes K_q B)
\end{aligned} \tag{9.28}$$

Proof: Let τ be an arbitrary permutation of $\{1,...,q\}$. We define

$$\sigma = \begin{pmatrix} 1 & 2 & \cdot & \cdot & \cdot & p & p+2 & \cdot & \cdot & \cdot & p+q \\ 1 & 2 & \cdot & \cdot & \cdot & p & p+\tau(1) & \cdot & \cdot & \cdot & p+\tau(q) \end{pmatrix}$$

Then $\varepsilon_\sigma = \varepsilon_\tau$ and

$$A \otimes T_\tau B = T_\sigma (A \otimes B)$$

Hence from (9.19) we have

$$K_{p+q}(A \otimes T_\sigma B) = K_{p+q}\left(T_\sigma (A \otimes B)\right) = \varepsilon_\sigma K_{p+q}(A \otimes B)$$
$$= \varepsilon_\sigma K_{p+q}(A \otimes B)$$

or, equivalently,

$$K_{p+q}(A \otimes \varepsilon_\tau T_\tau B) = K_{p+q}(A \otimes B) \tag{9.29}$$

Summing (9.29) overall τ, we obtain $(9.28)_1$, A similar argument implies (9.28).

In closing this section, we state without proof an expression for the skew-symmetric operator in terms of the components of its tensor argument. The formula is

$$K_r A = \frac{1}{r!} \delta^{i_1 \ldots i_r}_{j_1 \ldots j_r} A_{i_1 \ldots i_r} e^{j_1} \otimes \cdots \otimes e^{j_r} \tag{9.30}$$

We leave the proof of this formula as an exercise to the reader. A classical notation for the components of $K_r A$ is $A_{[j_1 \ldots j_r]}$, so from (9.29)

$$A_{[j_1 \ldots j_r]} = \frac{1}{r!} \delta^{i_1 \ldots i_r}_{j_1 \ldots j_r} A_{i_1 \ldots i_r} \tag{9.31}$$

Naturally, we call $K_r A$ the *skew-symmetric part* of A. The formula (9.30) imply that the component formula (8.69) defines a tensor K_r of order $2r$.

Exercises

1. Let $A \in T_r(V)$ and define an endomorphism S_r of $T_r(V)$ by

$$S_r A \equiv \frac{1}{r!} \sum_\sigma T_\sigma A$$

where the summation is taken overall permutations σ of $\{1,..., r\}$ as in (9.6). Naturally S_r is called the *symmetric operator*. Show that it is a projection from $T_r(V)$ into the subspace consisting of completely symmetric tensors of order r. What is the kernel $K(S_r)$? A classical notation for the components of S_rA is $A_{(j_1 \cdots j_r)}$.

2. Prove the formula (9.30).

Wedge Product

We have defined the concept of the tensor product $\otimes$, first for vectors, then generalised to tensors. We pointed out that the tensor product has the important universal factorisation property. In this section we shall define a similar operation, called the *wedge product* (or the *exterior product*), which we shall denote by the symbol $\wedge$. We shall define first the wedge product of any set of vectors.

If $(v^1,...,v^r)$ is any r-triple of vectors, then their tensor product $v^1 \otimes \cdots \otimes v^r$ is a simple tensor in $T_r(V)$. We have introduced the skew-symmetric operator K_r, which is a projection from $T_r(V)$ onto $\hat{T}_r(V)$. We now define

$$v^1 \wedge \cdots \wedge v^r \equiv \wedge(v^1,...,v^r) \equiv r!K_r(v^1 \otimes \cdots \otimes v^r) \tag{9.32}$$

for any vectors $v^1,...,v^r$. For example, if $r=2$, from (9.10) for the special case that $A = v^1 \otimes v^2$ we have

$$v^1 \wedge v^2 = v^1 \otimes v^2 - v^2 \otimes v^1 \tag{9.33}$$

In general, from (9.6) and (9.3) we have

$$\begin{aligned} v^1 \wedge \cdots \wedge v^r &= \sum_\sigma \varepsilon_\sigma T_\sigma (v^1 \otimes \cdots \otimes v^2) \\ &= \sum_\sigma \varepsilon_\sigma (v^{\sigma^{-1}(1)} \otimes \cdots \otimes v^{\sigma^{-1}(r)}) \\ &= \sum_\sigma \varepsilon_\sigma (v^{\sigma(1)} \otimes \cdots \otimes v^{\sigma(r)}) \end{aligned} \tag{9.34}$$

where in deriving (9.34), we have used the fact that

$$\varepsilon_\sigma = \varepsilon_{\sigma^{-1}} \tag{9.35}$$

and the fact that the summation is taken overall permutations σ of $\{1,..., r\}$.

We can regard $(9.32)_2$ as the definition of the operation

$$\wedge : \underbrace{V \times \cdots \times V}_{r \text{ times}} \rightarrow \hat{T}_p(V) \tag{9.36}$$

which is called the operation of *wedge product*. From $(9.32)_3$ it is clear that this operation, like the tensor product, is multilinear; further, $\wedge$ is a (completely) skew-symmetric operation in the sense that

$$\wedge(v^{\sigma(1)},...,v^{\sigma(r)}) = \varepsilon_\sigma \wedge (v^1,...,v^r) \tag{9.37}$$

for all $v^1,...,v^r \in V$ and all permutations σ of $\{1,..., r\}$. This fact follows directly from the skew symmetry of K_r. Next we show that the wedge product has also a universal factorisation property which is the condition asserted by the following.

Theorem 9.5: If W is an arbitrary completely skew-symmetric multilinear transformation

$$W : \underbrace{V \times \cdots \times V}_{r \text{ times}} \rightarrow U \tag{9.38}$$

where U is an arbitrary vector space, then there exists a unique linear transformation

$$D : \hat{T}_r(V) \rightarrow U \tag{9.39}$$

such that

$$W(v^1,...,v^r) = D(v^1 \wedge \cdots \wedge v^r) \tag{9.40}$$

for all $v^1,...,v^r \in V$. In operator form, (9.40) means that

$$W = D \circ \wedge \tag{9.41}$$

Proof: We can use the universal factorisation property of the tensor product to decompose W by

$$W = C \circ \otimes \tag{9.42}$$

where C is a linear transformation from $T_r(V)$ to U,

$$C : T_r(V) \to U \tag{9.43}$$

Now in view of the fact that W is skew-symmetric, we see that the particular linear transformation C has the property

$$C(v^{\sigma(1)} \otimes \cdots \otimes v^{\sigma(r)}) = \varepsilon_\sigma C(v^1 \otimes \cdots \otimes v^r) \tag{9.44}$$

for all simple tensors $v^1 \otimes \cdots \otimes v^r \in T_r(V)$ and all permutations σ of $\{1,...r\}$. Consequently, if we multiply (9.44) by ε_σ and sum the result overall σ, then from the linearity of C and the definition (9.34) we have

$$\begin{aligned} C(v^1 \wedge \cdots \wedge v^r) &= r!C(v^1 \otimes \cdots \otimes v^r) \\ &= r!W(v^1,...,v^r) \end{aligned} \tag{9.45}$$

for all $v^1,...,v^r \in V$. Thus the desired linear transformation D is simply given by

$$D = \frac{1}{r!} C_{\hat{T}_r(V)} \tag{9.46}$$

where the symbol on the right-hand denotes the restriction of C on $\hat{T}_r(V)$ as usual.

Uniqueness of D can be proved in exactly the same way as in the proof of Theorem 8.75. Here we need the fact that the tensors of the form $v^1 \wedge \cdots \wedge v^r \in \hat{T}_r(V)$, which may be called simply *skew-symmetric tensors* for an obvious reason, generate the space $\hat{T}_r(V)$. This fact is a direct consequence of the following results:

(*i*) $T_r(V)$ is generated by the set of all simple tensors $v^1 \otimes \cdots \otimes v^r$.

(*ii*) K_r is a linear transformation from $T_r(V)$ *onto* $\hat{T}_r(V)$.

(*iii*) Equation (9.32) which defines the simple skew-symmetric tensors $v^1 \wedge \cdots \wedge v^r$.

Knowing that the simple skew-symmetric tensors $v^1 \wedge \cdots \wedge v^r$ form a generating set of $\hat{T}_r(V)$, we can conclude immediately that the linear transformation D is unique, since its values on the generating set are uniquely determined by W through the basic condition (9.40).

As remarked before, the universal factorisation property can be used to define the tensor product abstractly. The preceding theorem shows that the same applies to the wedge product. In fact, by following this abstract approach, one can define the vector space $\hat{T}_r(V)$ entirely independent of the vector space $T_r(V)$ and the operation $\wedge$ entirely independent of the operation $\otimes$.

Having defined the wedge product for vectors, we can generalise the operation easily to skew-symmetric tensors. If $A \in \hat{T}_p(V)$ and $B \in \hat{T}_q(V)$, then we define

$$A \wedge B = \binom{r}{p} K_r(A \otimes B) \tag{9.47}$$

where, as before,

$$r = p + q \tag{9.48}$$

and

$$\binom{r}{p} = \frac{r(r-1)\cdots(r-p+1)}{p!} = \frac{r!}{p!q!} \tag{9.49}$$

We can regard (9.47) as the definition of the wedge product from $\hat{T}_p(V) \times \hat{T}_q(V) \to \hat{T}_q(V) \to \hat{T}_r(V)$,

$$\wedge : \hat{T}_p(V) \times \hat{T}_q(V) \to \hat{T}_r(V) \tag{9.50}$$

Clearly, this operation is bilinear and is characterised by the condition that

$$(v^1 \wedge \cdots \wedge v^p) \wedge (u^1 \wedge \cdots \wedge u^q) = v^1 \wedge \cdots \wedge v^p \wedge u^1 \wedge \cdots \wedge u^q \tag{9.51}$$

for all simple skew-symmetric tensors $v^1 \wedge \cdots \wedge v^p \in \hat{T}_p(V)$ and $u^1 \wedge \cdots \wedge u^q \in \hat{T}_q(V)$. We leave the proof of this simple fact as an exercise to the reader. In component form, the wedge product $A \wedge B$ is given by

$$A \wedge B = \frac{1}{p!q!} \delta^{i_1 \ldots i_r}_{j_1 \ldots j_r} A_{i_1 \ldots i_p} B_{i_{p+1} \ldots i_r} e^{j_1} \otimes \cdots \otimes e^{j_r} \tag{9.52}$$

relative to the product basis of any basis $\{e^j\}$ for V.

In view of (9.51), we can generalise the wedge product further to an arbitrary member of skew-symmetric tensors. For example, if $A \in \hat{T}_a(V), B \in \hat{T}_b(V)$ and $C \in \hat{T}_c(V)$, then we have

$$(A \wedge B) \wedge C = A \wedge (B \wedge C) = A \wedge B \wedge C \tag{9.53}$$

Moreover, $A \wedge B \wedge C$ is also given by

$$A \wedge B \wedge C = \frac{(a+b+c)!}{a!b!c!} K_{(a+b+c)}(A \otimes B \otimes C \otimes) \tag{9.54}$$

Further, the wedge product is multilinear in its arguments and is characterised by the associative law such as

$$\begin{aligned}(v^1 \wedge \cdots \wedge v^a) \wedge (u^1 \wedge \cdots \wedge u^b) \wedge (w^1 \wedge \cdots \wedge w^c)\\ v^1 \wedge \cdots \wedge v^a \wedge u^1 \wedge \cdots \wedge u^b \wedge w^1 \wedge \cdots \wedge w^c\end{aligned} \tag{9.55}$$

for all simple tensors involved.

From (9.55), the skew symmetry of the wedge product for vectors can be generalised to the condition such as

$$B \wedge A = (-1)^{pq} A \wedge B \tag{9.56}$$

for all $A \in \hat{T}_p(V)$ and $B \in \hat{T}_q(V)$. In particular, if A is of odd order, then

$$A \wedge A = 0 \tag{9.57}$$

since from (9.56) if the order p of A is odd, then

$$A \wedge A = (-1)^{p^2} A \wedge A = -A \wedge A \tag{9.58}$$

A special case of (9.57) is the elementary result that

$$v \wedge v = 0 \tag{9.59}$$

which is also obvious from (9.33).

Exercises

1. Verify (9.37), (9.51), $(9.53)_1$, and (9.54).
2. Show that (9.54) can be rewritten as

$$v^1 \wedge \cdots \wedge v^r = \delta^{1 \ldots r}_{i_1 \ldots i_r} v^{i_1} \otimes \cdots \otimes v^{i_r} \tag{9.60}$$

where the repeated indices are summed from 1 to r.

3. Show that

$$v^1 \wedge \cdots \wedge v^r(u_1, \ldots, u_r) = \det \begin{bmatrix} v^1 \cdot u_1 & \cdot & \cdot & \cdot & v^1 \cdot u_r \\ \cdot & & & & \cdot \\ \cdot & & & & \cdot \\ \cdot & & & & \cdot \\ v^r \cdot u_1 & \cdot & \cdot & \cdot & v^r \cdot u_r \end{bmatrix} \tag{9.61}$$

for all vectors involved.

4. Show that

$$e^{i_1} \wedge \cdots \wedge e^{i_r}(e_{j_1, \ldots,} e_{j_2}) = \delta^{i_1 \ldots i_r}_{j_1 \ldots j_r} \tag{9.62}$$

for any reciprocal bases $\{e^i\}$ and $\{e_i\}$.

Product Bases and Strict Components

We have remarked that the space $\hat{T}_r(V)$ is generated by the set of all simple skew-symmetric tensors of the form $v^1 \wedge \cdots \wedge v^r$. This generating set is linearly dependent, of course. In this section we shall determine a linearly independent generating set and thus a basis for $\hat{T}_r(V)$ consisting entirely of simple skew-symmetric tensors. Naturally, we call such a basis a basis a product basis for $\hat{T}_r(V)$.

Let $\{e^i\}$ be a basis for V as usual. Then the simple tensors $\{e^{i_1} \otimes \cdots \otimes e^{i_r}\}$ basis for $T_r(V)$. Since $\hat{T}_r(V)$, is a subspace of $T_r(V)$, every element $A \in \hat{T}_r(V)$ has the representation

$$A = A_{i_1 \ldots i_r} e^{i_1} \otimes \cdots \otimes e^{i_r} \tag{9.63}$$

where the repeated indices are summed from 1 to N, the dimension of $T_r(V)$. Now, since $A \in \hat{T}_r(V)$, it is invariant under the skew-symmetric operator K_r, namely

$$A = K_r A = A_{i_1 \ldots i_r} K_r(e^{i_1} \otimes \cdots \otimes e^{i_r}) \tag{9.64}$$

Then from (9.32) we can rewrite the representation (9.63) as

$$A = \frac{1}{r!} A_{i_1 \ldots i_r} e^{i_1} \wedge \cdots \wedge e^{i_r} \tag{9.65}$$

Thus we have shown that the set of simple skew-symmetric tensors $\{e^{i_1} \wedge \cdots \wedge e^{i_r}\}$ already forms a generating set for $\hat{T}_r(V)$.

The generating set $\{e^{i_1} \wedge \cdots \wedge e^{i_r}\}$ is still not linearly independent, however. This fact is easily seen, since the wedge product is skew-symmetric, as shown by (9.37). Indeed, if i_1 and i_2 are equal, then $e^{i_1} \wedge \cdots \wedge e^{i_r}$ must vanish. In general if σ is any permutation of $\{1,...r\}$, $e^{i_{\sigma(1)}} \wedge \cdots \wedge e^{i_{\sigma(r)}}$ then is linearly related to $e^{i_1} \wedge \cdots \wedge e^{i_r}$ by

$$e^{i_{\sigma(1)}} \wedge \cdots \wedge e^{i_{\sigma(r)}} = \varepsilon_\sigma e^{i_1} \wedge \cdots \wedge e^{i_r} \tag{9.66}$$

Hence, if we eliminate the redundant elements of the generating set $\{e^{i_1} \wedge \cdots \wedge e^{i_r}\}$ by restricting the range of the indices $(i_1,..., i_r)$ in such a way that

$$i_1 < \cdots < i_r \tag{9.67}$$

the resulting subset $\{e^{i_1} \wedge \cdots \wedge e^{i_r}, i_1 < \cdots < i_r\}$ remains a generating set of $\hat{T}_r(V)$. The next theorem shows that this subset is linearly independent and thus a basis for $\hat{T}_r(V)$, called the *product basis.*

Theorem 9.6: The set $\{e^{i_1} \wedge \cdots \wedge e^{i_r}, i_1 < \cdots < i_r\}$ is linearly independent.

Proof: Suppose that the set $\{e^{i_1} \wedge \cdots \wedge e^{i_r}, i_1 < \cdots < i_r\}$ obeys the homogeneous linear equation

$$\sum_{i_1<...<i_r} C_{i_1...i_r} e^{i_1} \wedge \cdots \wedge e^{i_r} \tag{9.68}$$

where $\{C_{i_1...i_r} i_1 < \cdots < i_r\}$ are scalars, and where the summation is taken overall indices $i_1,...,i_r$ from 1 to N subject to the condition (9.67). Then we must show that $C_{i_1...i_r}$ vanishes completely. From (9.62), if we evaluate the tensor (9.68) at the argument $(e_{i_1},...,e_{i_r})$, where $\{e_j\}$ denotes the reciprocal basis of $\{e^i\}$ as usual, we get

$$\sum_{i_1<...<i_r} C_{i_1...i_r} \delta^{i_1...i_r}_{j_1...j_r} = 0 \tag{9.69}$$

for all $j_1,...,j_r$ ranging from 1 to N. In particular, if we choose $j_1 < \cdots < j_r$, then the summation reduces to only one term, namely $C_{j_1 \cdots j_r}$ and the equation yields

$$C_{j_1 \cdots j_r} = 0$$

which is the desired result.

It is easy to see that there are only $\binom{N}{r}$ number of elements in the product basis

$$\{e^{i_1} \wedge \cdots \wedge e^{i_r}, i_1 < ... < i_r\} \tag{9.70}$$

Thus we have

$$\dim \hat{T}_r(V) = \binom{N}{r} = \frac{N!}{r!(N-r)!} \tag{9.71}$$

In particular, we recover the corollary of Theorem 9.6, that is

$$\dim \hat{T}_r(V) = 0$$

$$\text{if } r > N.$$

Returning now to the representation (9.65) for an arbitrary skew-symmetric tensor $A \in \hat{T}_r(V)$, we can rewrite that representation as

$$A = \sum_{i_1 < ... < i_r} A_{i_1 ... i_r} e^{i_1} \wedge \cdots \wedge e^{i_r} \tag{9.72}$$

The reason that we can replace the full summation in (9.65) by the restricted summation is because for each increasing r-tuple $(i_1,...,i_r)$ there are precisely $r!$ permutations of $\{i_1,...,i_r\}$. Further, their correspondingr terms in the full summation (9.65) are all equal to one another, since both $A_{i_1 ... i_r}$ and $e^{i_1} \wedge \cdots \wedge e^{i_r}$ are completely skew-symmetric in the indices $(i_1,...,i_r)$, so for any permutation σ of $\{i_1,...,i_r\}$ we have

$$A_{i_{\sigma(1)} \cdots i_{\sigma(r)}} e^{i_{\sigma(1)}} \wedge \cdots \wedge e^{i_{\sigma(r)}} = \varepsilon_\sigma^2 A_{i_1 ... i_r} e^{i_1} \wedge \cdots \wedge e^{i_r} = A_{i_1 ... i_r} e^{i_i} \wedge \cdots \wedge e^{i_r}$$

In view of (9.72), we see that the scalars

$$\{A_{i_1 \dots i_r}, i_1 < \cdots < i_r\}$$

are the components of A relative to the product basis (9.70). For definiteness, we call these scalars the *strict components* of A.

As an illustration of this concept, let us compute the strict components of the simple skew-symmetric tensor $v^1 \wedge \cdots \wedge v^r \in \hat{T}_r(V)$. As usual we represent the vector v^i in component form relative to $\{e^j\}$ by

$$v^1 = v^i_j e^j$$

for all $i = 1,\dots,r$. Using the skew-symmetry of the wedge product, we have

$$v^1 \wedge \cdots \wedge v^r = v^1_{j_1} \cdots v^r_{j_r} e^{j_1} \wedge \cdots \wedge e^{j_r} = \sum_{i_1 < \dots < i_r} v^1_{j_1} \cdots v^r_{j_r} \delta^{j_1 \dots j_r}_{i_1 \dots i_r} e^{i_1} \wedge \cdots \wedge e^{i_r} \tag{9.73}$$

which means that the strict components of $v^1 \wedge \cdots \wedge v^r$ are

$$\left\{ v^1_{j_1} \cdots v^r_{j_r} \delta^{j_1 \dots j_r}_{i_1 \dots i_r}, i_1 < \cdots < i_r \right\}$$

Next, we consider the transformation rule for the strict components in general.

Theorem 9.7: Under a change of basis from $\{e^j\}$ to $\{\hat{e}^j\}$, the strict components of a tensor $A \in \hat{T}_r(V)$ obey the transformation rule

$$\hat{A}_{j_1 \dots j_r} = \sum_{i_1 < \dots < i_r} T^{i_1 \dots i_r}_{j_1 \dots j_r} A_{i_1 \dots i_r} \tag{9.74}$$

where $T^{i_1 \dots i_r}_{j_1 \dots j_r}$ is an $r \times r$ minor of the transformation matrix $[T^i_j]$ as defined by (6.37). Of course, T^i_j is given by

$$T^i_j = e^i \cdot e_j$$

as usual, where $\{\hat{e}_j\}$ is the reciprocal basis of $\{\hat{e}^j\}$.

Proof: Since the strict components are nothing but the ordinary tensor components restricted to the subset of indices in increasing order, we can compute their values in the usual way by

$$\hat{A}_{j_1 \ldots j_r} = A(\hat{e}_{j_1}, \ldots, \hat{e}_{j_1}) \tag{9.75}$$

Substituting the strict component representation (9.72) into (9.75) and making use of the formula (9.61) we obtain

$$\hat{A}_{j_1 \ldots j_r} = \sum_{i_1 < \ldots < i_r} A_{i_1 \ldots i_r} e^{i_1} \wedge \cdots \wedge e^{i_r} (\hat{e}_{j_1}, \ldots, \hat{e}_{j_r})$$

$$= \sum_{i_1 < \ldots < i_r} A_{i_1 \ldots i_r} \det \begin{bmatrix} e^{i_1} \cdot \hat{e}_{j_1} & \cdot & \cdot & \cdot & e^{i_1} \cdot \hat{e}_{j_r} \\ \cdot & & & & \cdot \\ \cdot & & & & \cdot \\ \cdot & & & & \cdot \\ e^{i_r} \cdot \hat{e}_{j_1} & & & & e^{i_r} \cdot \hat{e}_{j_r} \end{bmatrix}$$

Which is the desired result

From (6.41), we can write the transformation rule (9.74) in the form

$$\hat{A}_{j_1 \ldots j_r} = \det[T_b^a] \sum_{i_1 < \ldots < i_r} \left(\operatorname{cof} \hat{T}_{i_1 \ldots i_r}^{j_1 \ldots j_r} \right) A_{i_1 \ldots i_r} \tag{9.76}$$

As all illustration of (9.74), take $r = N$. Then (9.71) implies

$$\dim \hat{T}_N(V) = 1 \tag{9.77}$$

And (9.74) reduces to

$$\hat{A}_{12 \ldots N} = \det\left[T_j^i \right] A_{12 \ldots N} \tag{9.78}$$

Comparing (9.78) with (8.52), we see that the strict component of an N-vector transforms according to the rule of an axial scalar of weight 1. *N*-vectors are also called *densities* or *density tensors.*

Next take $r = N-1$. Then (9.71) implies that

$$\dim \hat{T}_{N-1}(V) = N \tag{9.79}$$

In this case the transformation rule (9.74) can be rewritten as

$$\hat{A}^k = \det\left[T_j^i \right] \hat{T}_l^k A^l \tag{9.80}$$

where the quantities $\hat{A}^k$ and A^l are defined by

$$\hat{A}^k \equiv (-1)^{N-k} A_{12\ldots\breve{k}\ldots N} \tag{9.81}$$

and similarly

$$A^l \equiv (-1)^{N-l} A_{12\ldots\breve{l}\ldots N} \tag{9.82}$$

Here the symbol $\breve{}$ over k or l means k or l are deleted from the list of indices as before. To prove (9.80), we make use of the alternative form (9.76), obtaining

$$\hat{A}_{12\ldots\breve{k}\ldots N} = \det\left[T_b^a\right]\sum_l\left(\operatorname{cof}\hat{T}_{12\ldots\breve{l}\ldots N}^{12\ldots\breve{k}\ldots N}\right)A_{12\ldots\breve{l}\ldots N}$$

Multiplying this equation by $(-1)^{N-k}$ and using the definitions (9.81) and (9.82), we get

$$\hat{A}^k = \det\left[T_b^a\right]\sum_l(-1)^{k+1}\left(\operatorname{cof}\hat{T}_{12\ldots\breve{l}\ldots N}^{12\ldots\breve{k}\ldots N}\right)A^l$$

But now from (6.39), it is easy to see that the cofactor of the $(N-1)$ $(N-1)$ minor $\hat{T}_{12\ldots\breve{l}\ldots N}^{12\ldots\breve{k}\ldots N}$ in the matrix $\left[\hat{T}_b^a\right]$ is simply $(-1)^{k+1}$ times the element $\hat{T}_l^k$. Thus (9.80) is proved.

From (9.81) or (9.82) we see that the quantities $\hat{A}^k$ and A^l, like the strict components $\hat{A}_{12\ldots\breve{k}\ldots N}$ and $A_{12\ldots\breve{l}\ldots N}$ characterise the tensor A completely. In view of the transformation rule (9.80), we see that the quantity A^l transforms according to the rule of the component of an axial vector of weight 1. $(N-1)$ vectors are often called *(axial) vector densities.*

The operation given by (9.81) and (9.82) can be generalised to $\hat{T}_{N-r}(V)$ in general. If $A_{i_1\ldots i_{N-r}}$ is a strict component of $A \in \breve{T}_{N-r}(V)$, then we define a quantity $A^{j_1\cdots j_r}$ by

$$\begin{aligned} A^{j_1\cdots j_r} &= \sum_{i_1<\ldots<i_{N-r}} \varepsilon^{i_1\cdots i_{N-r}j_1\cdots j_r} A_{i_1\ldots i_{N-r}} \\ &= \frac{1}{(N-r)!}\varepsilon^{i_1\cdots i_{N-r}j_1\cdots j_r} A_{i_1\ldots i_{N-r}} \end{aligned} \tag{9.83}$$

When $r=1$, (9.83) reduces to (9.82). Recall that $\varepsilon^{i_1\cdots i_N}$ transforms according to the rule of an axial contravariant tensor of order N

and weight 1. Moreover, since $A^{j_1 \cdots j_r}$ is skew-symmetric in $(j_1, \ldots, j_r)$, we can write the transformation rule as

$$\hat{A}^{j_1 \cdots j_r} = \det[T_b^a] \sum_{i_1 < \ldots < i_r} \hat{T}_{i_1 \ldots i_r}^{j_1 \cdots j_r} A^{i_1 \ldots i_r} \tag{9.84}$$

where $\hat{T}_{i_1 \ldots i_r}^{j_1 \cdots j_r}$ is an $r \times r$ minor of the transformation matrix $[T_b^a]$. When $r = 1$, (9.84) reduces to (9.80).

As an illustration of (9.82), we take $N = 3$; then (9.82) yields

$$A^1 = A_{23}, \qquad A^2 = -A_{13}, \qquad A^3 = A_{12} \tag{9.85}$$

As we shall see, this operation is closely related to the classical operation of *cross product* (or vector product) which assigns an axial vector to a skew- symmetric two-vector on a three-dimensional space.

Before closing this section, we note here a convenient condition for determining whether or not a given set of vectors is linearly dependent by examining their wedge product.

Theorem 9.8: A set of r vectors $\{v^1, \ldots v^r\}$ is linearly dependent if and only if their wedge product vanishes, i.e.,

$$v^1 \wedge \cdots \wedge v^r = 0 \tag{9.86}$$

Proof: Necessity is obvious, since if $\{v^1, \ldots v^r\}$ is linearly dependent, then at least one of the vectors can be expressed as a linear combination of the other vectors, say

$$v^r = \alpha_1 v^1 + \cdots + \alpha_{r-1} v^{r-1} \tag{9.87}$$

Then (9.86) follows because

$$v^1 \wedge \cdots \wedge v^r = \sum_{j=1}^{r-1} \alpha_j v^1 \wedge \cdots \wedge v^{r-1} \wedge v^j = 0$$

as required by the skew symmetry of the wedge product. Conversely if $\{v^1, \ldots, v^r\}$ is linearly independent, then it can be extended to a basis $\{v^1, \ldots, v^N\}$. From Theorem 9.6, the simple skew-symmetric tensor $v^1 \wedge \cdots \wedge v^N$ forms a basis for the one-dimensional space $\hat{T}_N(V)$ and thus is non-zero, so that this factor $v^1 \wedge \cdots \wedge v^r$ is also non-zero.

Corollary: A set of N covectors $\{v^1,...,v^N\}$ is a basis for V if and only if $v^1 \wedge \cdots \wedge v^N \neq 0$.

Exercises

1. Show that the product basis of $\hat{T}_r(V)$ satisfies the following transformation rule:

$$e^{i_1} \wedge \cdots \wedge e^{i_r} = \sum_{j_1<\ldots<j_r} T^{i_1\ldots i_r}_{j_1\ldots j_r} \hat{e}^{j_1} \wedge \cdots \wedge \hat{e}^{j_r} \tag{9.88}$$

   relative to any change of basis from $\{\hat{e}^i\}$ to $\{\hat{e}^j\}$ with transformation matrix $\left[T^i_j\right]$.

2. For the skew-symmetric tensor space $\hat{T}_r(V)$ it is customary to define the inner produce

$$* : \hat{T}_r(V) \times \hat{T}_r(V) \to R \tag{9.89}$$

   by requiring the product basis of an orthonormal basis be orthonormal with respect to *. In other words, relative to an orthonormal basis $\{i^1,...,i^N\}$ the inner produce $A*B$ of any $A, B \in \hat{T}_r(V)$ is given by

$$A * B = \sum_{i_1<\cdots<i_r} A_{i_1\ldots i_r} B_{i_1\ldots i_r} \tag{9.90}$$

   Show that this inner product is related to the ordinary inner product . on $\hat{T}_r(V)$, regarded as a subspace of $T_r(V)$, by

$$A * B = \frac{1}{r!} A.B \tag{9.91}$$

   for all $A, B \in \hat{T}_r(V)$.

3. Relative to the inner product $*$, show that

$$(v^1 \wedge \cdots \wedge v^r) * (u^1 \wedge \cdots \wedge u^r) = \det \begin{bmatrix} v^1 \cdot u^1 & \cdot & \cdot & \cdot & v^1 \cdot u^r \\ \cdot & & & & \cdot \\ \cdot & & & & \cdot \\ \cdot & & & & \cdot \\ v^r \cdot u^1 & \cdot & \cdot & \cdot & v^r \cdot u^r \end{bmatrix} \tag{9.92}$$

for all simple skew- symmetric tensors $v^1 \wedge \cdots \wedge v^r$ and $u^1 \wedge \cdots \wedge u^r$ in $\hat{T}_r(V)$.

4. Relative to the product basis of any basis $\{e^i\}$, show that the inner product A $*$ B has the representation

$$A * B = \sum_{\substack{i_1<\ldots<i_r \\ j_1<\ldots<j_r}} \det \begin{bmatrix} e^{i_1j_1} & \cdot & \cdot & \cdot & e^{i_1j_r} \\ \cdot & & & & \cdot \\ \cdot & & & & \cdot \\ \cdot & & & & \cdot \\ e^{i_rj_1} & \cdot & \cdot & \cdot & e^{i_rj_r} \end{bmatrix} A_{i_1\ldots i_r} B_{j_1\ldots j_r} \tag{9.93}$$

where e^{ij} is given by

$$e^{ij} = e^i \cdot e^j \tag{9.94}$$

as before. In particular, if $\{e^i\}$ is orthonormal, then $e^{ij} = \delta^{ij}$ and (9.93) reduces to (9.90).

5. Prove the transformation rule (9.84).

Determinant and Orientation

We have shown that the strict component of an N-vector over an N-dimensional space V transforms according to the rule of an axial scalar of weight 1. For brevity, we call an N-vector simply a *density* or a *density tensor*. From (9.77), the space $\hat{T}_N(V)$ of densities on V is one-dimensional, and for any basis $\{e^i\}$ a product basis for $\hat{T}_N(V)$ is $e^1 \wedge \cdots \wedge e^N$. If D is a density, then it has the representation

$$D = D_{12\ldots N} e^1 \wedge \cdots \wedge e^N \tag{9.95}$$

where the strict component $D_{12\ldots N}$ is given by

$$D_{12\ldots N} = D(e_1, \ldots, e_N) \tag{9.96}$$

$\{e_i\}$ being the reciprocal basis of $\{e^j\}$ as usual. The representation (9.95) shows that every density is a simple skew-symmetric tensor, e.g., we can represent D by

$$D = (D_{12\ldots N} e^1) \wedge \cdots \wedge e^N \tag{9.97}$$

This representation is not unique, of course.

Now if A is an endomorphism of V, we can define a linear map

$$f : \hat{T}_N(V) \to \hat{T}_N(V) \tag{9.98}$$

by the condition

$$f(v^1 \wedge \cdots \wedge v^N) = (Av^1) \wedge \cdots \wedge (Av^N) \tag{9.99}$$

for all $v^1, \ldots, v^N \in V$. We can prove the existence and uniqueness of the linear map f by the universal factorisation property of $\hat{T}_N(V)$ as shown by Theorem 9.5. Indeed, we define first the skew-symmetric multilinear map

$$F : V \times \cdots \times V \to \hat{T}_N(V) \tag{9.100}$$

by

$$F(v^1, \ldots, v^N) = (Av^1) \wedge \cdots \wedge (Av^N) \tag{9.101}$$

where the skew symmetry and the multilinearity of F are obvious. Then from (9.41) we can define a unique linear map f of the form (9.98) such that

$$F = f \circ \wedge \tag{9.102}$$

which means precisely the condition (9.99).

Since $\hat{T}_N(V)$ is one-dimensional, the linear map f must have the representation

$$f(v^1 \wedge \cdots \wedge v^N) = \alpha v^1 \wedge \cdots \wedge v^N \tag{9.103}$$

where α is a scalar uniquely determined by f and hence A. We claim that

$$\alpha = \det A \tag{9.104}$$

Thus the determinant of A can be defined free of any basis by the condition

$$(Av^1) \wedge \cdots \wedge (Av^N) = (\det A) v^1 \wedge \cdots \wedge v^N \tag{9.105}$$

for all $v^1, \ldots, v^N \in V$.

To prove (9.104), or, equivalently, (9.105), we choose reciprocal basis $\{e_i\}$ and $\{e^i\}$ for V and recall that the determinant of A is given by the determinant of the component matrix of A. Let A be represented by the component forms

$$Ae_i = A^j{}_i e_j, \qquad Ae^i = A^i_j e^j \tag{9.106}$$

where $(9.106)_2$ is meaningful because V is an inner product space. Then, by definition, we have

$$\det A = \det[A^j_i] = \det[A^i_j] \tag{9.107}$$

Substituting $(9.106)_2$ into (9.99), we get

$$f(e^1 \wedge \cdots \wedge e^N) = A_{ji}{}^1 \ldots A_{j_N}{}^N e^{ji} \wedge \cdots \wedge e^{j_N} \tag{9.108}$$

The skew symmetry of the wedge product implies that

$$e^{j_1} \wedge \cdots \wedge e^{j_N} = \varepsilon^{j_1 \ldots j_N} e^1 \wedge \cdots \wedge e^N \tag{9.109}$$

where the ε symbol is defined by (6.3). Combining (9.108) and (9.109), and comparing the result with (9.103), we see that

$$\alpha e^1 \wedge \cdots \wedge e^N = A_{j_1}{}^1 \cdots A_{j_N}{}^N \varepsilon^{j_1,\ldots,j_N} e^1 \wedge \cdots \wedge e^N \tag{9.110}$$

or equivalently

$$\alpha = A_{j_1}{}^1 \ldots A_{j_N}{}^N \varepsilon^{j_1,\ldots,j_N} \tag{9.111}$$

since $e^1 \wedge \cdots \wedge e^N$ is a basis for $\hat{T}_N(V)$. But by definition the right-hand side of (9.111) is equal to the determinant of the matrix $[A^i_j]$ and thus (9.104) is proved.

As an illustration of (9.105), we see that

$$\det I = 1 \tag{9.112}$$

since we have

$$(Iv^1) \wedge \cdots \wedge (Iv^N) = v^1 \wedge \cdots \wedge v^N = (\det I) v^1 \wedge \cdots \wedge v^N$$

More generally, we have

$$\det(\alpha I) = \alpha^N \tag{9.113}$$

Since

$$(\alpha Iv^1)\wedge\cdots\wedge(\alpha Iv^N)=(\alpha v^1)\wedge\cdots\wedge(\alpha v^N)=\alpha^N v^1\wedge\cdots\wedge v^N$$

It is now also obvious that

$$\det(AB)=(\det A)(\det B) \tag{9.114}$$

since we have

$$\begin{aligned}(ABv^1)\wedge\cdots\wedge(ABv^N)&=(\det A)(Bv^1)\wedge\cdots\wedge(Bv^N)\\&=(\det A)(\det B)v^1\wedge\cdots\wedge v^N\end{aligned}$$

for all $v^1,\ldots,v^N\in V$. Combining (9.113) and (9.114), we recover also the formula

$$\det(\alpha A)=\alpha^N(\det A) \tag{9.115}$$

Another application of (9.105) yields the result that A is non-singular if and only if det A is non-zero. To see this result, notice first the simple fact that A is non-singular if and only if A transforms any basis $\{e^j\}$ of V into a basis $\{Ae^j\}$ of V. From the corollary of Theorem 9.8, $\{Ae^j\}$ is a basis of V if and only if $(Ae^1)\wedge\cdots\wedge(Ae^N)\neq 0$. Then from (9.105), $(Ae^1)\wedge\cdots\wedge(Ae^N)\neq 0$ if and only if det $A\neq 0$. Thus det $A\neq 0$ is necessary and sufficient for A to be non-singular.

The determinant, of course, is just one of the invariants of A. We have introduced the set of fundamental invariants $\{\mu_1,\ldots,\mu_N\}$ for A by the equation

$$\det(A+tI)=t^N+\mu_1 t^{N-1}+\cdots+\mu_{N-1}t+\mu_N \tag{9.116}$$

Since we have shown that the determinant of any endomorphism can be characterised by the condition (9.105), if we apply (9.105) to the endomorphism $A + tI$, the result is

$$\begin{aligned}&(Av^1+tv^1)\wedge\cdots\wedge(Av^N+tv^N)\\&=(t^N+\mu_1 t^{N-1}+\cdots+\mu_{N-1}t+\mu_N)v^1\wedge\cdots\wedge v^N\end{aligned}$$

Comparing the coefficient of t^{N-k} on the two sides of this equation, we obtain

$$\mu_k v^1 \wedge \cdots \wedge v^N$$
$$= \sum_{i_1 < \cdots < i_K} v^1 \wedge \cdots \wedge Av^{i_1} \wedge \cdots \wedge Av^{i_2} \wedge \cdots \wedge Av^{i_K} \wedge \cdots \wedge v^N \quad (9.117)$$

where there are precisely $\binom{N}{K}$ terms in the summation on the right hand side of (9.117), each containing K factors of A acting on the covectors $v^{i_1},\ldots,v^{i_K}$. In particular, taking $K = N$, recover the result

$$\mu_N v^1 \wedge \cdots \wedge v^N = (Av^1) \wedge \cdots \wedge (Av^N)$$

which is the same as (9.105) since

$$\mu_N = \det A$$

Likewise, taking $K = 1$, we obtain

$$\mu_1 v^1 \wedge \cdots \wedge v^N = \sum_{i=1}^{N} v^1 \wedge \cdots \wedge Av^i \wedge \cdots \wedge v^N \quad (9.118)$$

which characterises the trace of A free of any basis, since

$$\mu_1 = trA$$

Of course, we can prove the existence and the uniqueness of μ_K satisfying the condition (9.117) by the universal factorisation property as before.

Since the space $\hat{T}_N(V)$ is one-dimensional, it is divided by 0 into two non-zero segments. Two non-zero densities D_1 and D_2 are in the same segment if they differ by a positive scalar factor, say $D_2 = \lambda D_1$, where $\lambda > 0$. Conversely, if D_1 and D_2 differ by a negative scalar factor, then they belong to different segments. If $\{e^i\}$ and $\{\hat{e}^i\}$ are two basis for V, we define their corresponding product basis $e^1 \wedge \cdots \wedge e^N$ and $\hat{e}^1 \wedge \cdots \wedge \hat{e}^N$ for $\hat{T}_N(V)$ as usual; then we say that $\{e^i\}$ and $\{\hat{e}^i\}$ have the *same orientation* and that the change of basis is a *proper transformation* if $e^1 \wedge \cdots \wedge e^N$ and $\hat{e}^1 \wedge \cdots \wedge \hat{e}^N$ belong to the same segment of $\hat{T}_N(V)$. Conversely, *opposite orientation* and *improper transformation* are defined by the condition that $e^1 \wedge \cdots \wedge e^N$ and $\hat{e}^1 \wedge \cdots \wedge \hat{e}^N$ belong to the different segments of $\hat{T}_N(V)$. From (9.88) for the case $r = N$, the product basis $e^1 \wedge \cdots \wedge e^N$ and $\hat{e}^1 \wedge \cdots \wedge \hat{e}^N$ are related by

$$e^1 \wedge \cdots \wedge e^N = \det[T^i_j]\hat{e}^1 \wedge \cdots \wedge \hat{e}^N \tag{9.119}$$

Consequently, the change of basis is proper or improper if and only if det $[T^i_j]$ is positive or negative, respectively.

It is conventional to designate one segment of $\hat{T}_N(V)$ *positive* and the other one *negative*. Whenever such a designation has been made, we say that $\hat{T}_N(V)$ and, thus, V, are *oriented*. For an oriented space V a basis $\{e^i\}$ is *positively oriented* or *right-handed* if its product basis $e^1 \wedge \cdots \wedge e^N$ belongs to the positive segment of $\hat{T}_N(V)$; otherwise, the basis is *negatively oriented* or *left-handed*. It is customary to restrict the choice of basis for an oriented space to positively oriented bases only. Under such a restriction, the parity ε of all relative tensors always has the value +1, since the transformation is restricted to be proper. Hence, in this case it is not necessary to distinguish relative tensors into axial ones and polar ones.

It is conventional to use the inner product $*$ defined by (9.91) for skew-symmetric tensors. For the space of densities $\hat{T}_N(V)$ the inner product is given by

$$A * B = \det[e^{ij}]A_{12\ldots N}B_{12\ldots N} \tag{9.120}$$

where $A_{12\ldots N}$ and $B_{12\ldots N}$ are the strict component of A and B as defined by (9.96). Clearly, there exists an unique unit density with respect to $*$ in each segment of $\hat{T}_N(V)$. If $\hat{T}_N(V)$ is oriented, the unit density in the positive segment is usually denoted by E, then the unit density in the negative segment is $-E$. In general, if D is any density in $\hat{T}_N(V)$, then it has the representation

$$D = dE \tag{9.121}$$

In accordance with the usual practice, the scalar d in (9.122), which is the component of D relative to the positive unit density E, is also called a density or more specifically a density scalar, as opposed to the term *density tensor* for D. This convention is consistent with the common practice of identifying a scalar α with an element $\alpha 1$ in R, where R is, of course, an oriented one-dimensional space with positive unit element 1.

If $\{e^i\}$ is a basis in an oriented space V, then we define the *density* e^* of $\{e^i\}$ to be the component of the product basis $e^1 \wedge \cdots \wedge e^N$ relative to E, namely

$$e^1 \wedge \cdots \wedge e^N = e^* E \tag{9.122}$$

In other words, e^* is the density of the product basis of $\{e^i\}$. Clearly, $\{e^i\}$ is positively oriented or negatively oriented depending on whether its density e^* is positive or negative, respectively. We say that a basis $\{e^i\}$ is *unimodular* if its density has unit absolute value, i.e., $|e^*| = 1$. All orthonormal bases, right-handed or left-handed, are always unimodular. A unimodular basis in general need not be orthonormal, however.

From (9.122) and (9.120), we have

$$e^{*^2} = e^* E * e^* E = \det[e^{ij}] \tag{9.123}$$

for any basis $\{e^i\}$. Hence the *absolute density* $|e^*|$ of $\{e^i\}$ is given by

$$|e^*| = \left(\det[e^{ij}]\right)^{1/2} \tag{9.124}$$

Substituting (9.124) into (9.122), we have

$$e^1 \wedge \cdots \wedge e^N = \varepsilon\left(\det[e^{ij}]^{1/2}\right) E \tag{9.125}$$

where ε is +1 if $\{e^i\}$ is positively oriented and it is –1 if $\{e^i\}$ negatively oriented.

From (9.119) and (9.121) the density of a basis transforms according to the rule

$$e^* = \det[T^i{}_j]\hat{e}^* \tag{9.126}$$

under a change of basis from $\{e^j\}$ to $\{\hat{e}^j\}$. Comparing (9.126) with (8.72), we see that the density of a basis is an axial relative scalar of weight –1.

An interesting property of a unit density (E or $-E$) is given by the following theorem.

Theorem 9.9: If U is a unit density in $\hat{T}_N(V)$, then

$$\left[U*(v^1\wedge\cdots\wedge v^N)\right]\left[U*(u^1\wedge\cdots\wedge u^N)\right]=\det\left[v^i\cdot u^j\right] \tag{9.127}$$

for all $v^1,\ldots,v^N$ and $u^1,\ldots,u^N$ in V.

Proof: Clearly we can represent U as the product basis of an orthonormal basis, say $\{i_k\}$, with

$$U=i_1\wedge\cdots\wedge i_N \tag{9.128}$$

From (9.128) and (9.92), we then have

$$U*(v^1\wedge\cdots\wedge v^N)=\det[i_j\cdot v^i]=\det[v^i{}_j]$$

and

$$U*(u^1\wedge\cdots\wedge u^N)=\det[i_j\cdot u^i]=\det[u^i{}_j]$$

where $v^i{}_j$ and $u^i{}_j$ are the j th components of v^i and u^i relative to $\{i_k\}$,respectively. Using the product rule and the transpose rule of the determinant, we then obtain

$$\begin{aligned}
&[U*(v^1\wedge\cdots\wedge v^N)][U*(u^1\wedge\cdots\wedge u^N)]\\
&=\det[v^i{}_j]\det[u^k{}_l]\\
&=\det[v^i{}_j]\det[u^k{}_l]^T=\det([v^i{}_j]\det[u^k{}_l]^T)\\
&=\det\left[\sum_{k=1}^{N}v^i{}_k u^i{}_k\right]=\det[v^i\cdot u^j]
\end{aligned}$$

which is the desired result.

By exactly the same argument, we have also the following theorem.

Theorem 9.10: If U is a unit density as before, then

$$U\ (v_1,...,v_N)\ U\ (u_1,...,u_N) = \det\ [v_i\ .\ u_j] \tag{9.129}$$

for all $v_1,...,v_N$ and $v_1,...,v_N$ in V.

Exercises

1. If A is an endomorphism of V show that the determinant of A can be characterised by the following basis-free condition:

$$D\ (Av_1,...,\ Av_N) = (\det A)\ D\ (v_1,...,v_N) \tag{9.130}$$

for all densities $D \in \hat{T}_N(V)$ and all vectors $v_1,...,v_N$ in V.

Note.

A complete proof of this result consists of the following two parts:

(*i*) There exists a unique scalar α, depending on A, such that

$$D(Av_1,...,Av_N) = \alpha D(v_1,...,v_N)$$

for all $D \in \hat{T}_N(V)$ and all vectors $v_1,...v_N$ in V.

(*ii*) The scalar α is equal to det A.

2. Use the formula (9.130) and show that the fundamental invariants $\{\mu_1,...\mu_N\}$ endomorphism A can be characterised by

$$\mu_k D(v_1,...,v_N) = \sum_{i_1<...<i_K} D(v_1,...,Av_{i_1},...,Av_{i_K},...,v_N) \tag{9.131}$$

for all $v_1,...,v_N \in V$ and all $D \in \hat{T}_N(V)$. Here the summation on the right-hand side of (9.131) is similar to that of (9.117), i.e., there are $\binom{N}{K}$ terms in the summation, each containing the value of D at the argument with K vectors $v_{i_1},...,v_{i_K}$, acted on by A. In particular, taking $K = N$, we recover the formula (9.130), and taking $K = 1$, we obtain

$$(trA)D(v_1,...,v_N) = \sum_{i=1}^{N} D(v_1,...,Av_i,...,v_N) \tag{9.132}$$

3. The *Gramian* is a function (not multilinear) G of K vectors defined by

$$G(v_1,...,v_N) = \det[v_i \cdot v_j] \tag{9.133}$$

where the matrix $[v_i \cdot v_j]$ is $K \times K$, of course, Use the result of Theorem 9.10 and show that

$$G(v_1,...,v_N] \geq 0 \tag{9.134}$$

and that the equality holds if and only if $\{v_1,...,v_K\}$ is a linear dependent set. *Note.* When $K = 2$, the result (9.134) reduces to the Schwarz inequality.

4. Use the results of this section and prove that

   $$\det A = \det A^T$$

   for an endomorphism $A \in L(V;V)$.

5. Show that adj A, which is used in (7.26), can be defined by

   $$(\operatorname{adj} A)v^1 \wedge v^2 \wedge \cdots \wedge v^N = v^1 \wedge Av^2 \wedge \cdots \wedge Av^N$$

6. Use (9.117) and the result in preceding exercise 9.99 and show that

   $$\mu_{N-1} = \operatorname{tr} \operatorname{adj} A$$

7. Show that

   adj AB = adj B adj A

   det adj A = $(\det A)^{N-1}$

   adj (adj A) = $(\det A)^{N-2} A$

and

det adj (adj A) = $(\det A)^{(N-1)2}$

for $A, B \in L(V;V)$ and $N = \dim V$.

The Duality

As a consequence of (9.71), the dimension of the space $\hat{T}_r(V)$ is equal to that of $\hat{T}_{N-r}(V)$. Hence the spaces $\hat{T}_r(V)$ and $\hat{T}_{N-r}(V)$ are isomorphic. The purpose of this section is to establish a particular isomorphism,

$$D_r : \hat{T}_r(V) \to \hat{T}_{N-r}(V) \tag{9.135}$$

called the *duality operator*, for an oriented space V. As we shall see, this duality operator gives rise to a definition to the operation of *cross product* or *vector product* when $N = 3$ and $r = 2$.

We recall first that $\hat{T}_r(V)$ is equipped with the inner product $*$ given by (9.91). Let E be the distinguished positive unit density in $\hat{T}_N(V)$. Then for any $A \in \hat{T}_r(V)$ we define its *dual* D_rA in $\hat{T}_{N-r}(V)$ by the condition

$$E * (A \wedge Z) = (D_r A) * Z \tag{9.136}$$

for all $Z \in \hat{T}_{N-r}(V)$. Since the left-hand side of (9.136) is linear in Z, and since $*$ is an inner product, D_rA is uniquely determined by (9.136). Further, from (9.136), D_rA depends linearly on A, so D_r, is a linear transformation from $\hat{T}_r(V)$ to $\hat{T}_{N-r}(V)$.

Theorem 9.11: The duality operator D_r defined by the condition (9.136) is an isomorphism.

Proof: Since $\hat{T}_r(V)$ and $\hat{T}_{N-r}(V)$ are isomorphic, it suffices to show that D_r is an isometry, i.e.,

$$(D_r A) * (D_r A) = A * A \tag{9.137}$$

for all $A \in \hat{T}_r(V)$. The polar identity

$$A * B = \frac{1}{2}\{A * A + B * B - (A - B) * (A - B)\} \tag{9.138}$$

then implies that D_r preserves the inner product also; i.e.,

$$(D_r A) * (D_r B) = A * B \tag{9.139}$$

for all A and B in $\hat{T}_r(V)$. To prove (9.137), we choose an arbitrary right-handed orthogonal basis $\{i_j\}$ of V. Then E is simply the product basis of $\{i_j\}$

$$E = i_1 \wedge \cdots \wedge i_N \tag{9.140}$$

Of course, we represent A, Z, and, D_rA in strict component form relative to $\{i_j\}$ also; then (9.136) can be written as

$$\sum_{\substack{i_1<\ldots<i_r \\ j_1<\ldots<j_{N-r}}} \varepsilon_{i_1\ldots i_r j_1\ldots j_{N-r}} A_{i_1\ldots i_r} Z_{j_1\ldots j_{N-r}}$$

$$= \sum_{i_1<\ldots<j_{N-r}} (D_r A)_{j_1\ldots j_{N-r}} A_{i_1\ldots i_r} Z_{j_1\ldots j_{N-r}} \tag{9.141}$$

Since Z is arbitrary, (9.141) implies

$$(D_r A)_{j_1\ldots j_{N-r}} = \sum_{i_1<\ldots<j_r} \varepsilon_{i_1\ldots i_r j_1\ldots j_{N-r}} A_{i_1\ldots i_r} \tag{9.142}$$

This formula gives the strict components of $D_r A$ in terms of those of A relative to a right-handed orthonormal basis.

From (9.142) we can compute the value of left-hand side of (9.137) by

$$(D_r A) * (D_r A) = \sum_{\substack{i_1<\ldots<i_r \\ k_1<\ldots<k_r \\ j_1<\ldots<j_{N-r}}} \varepsilon_{i_1\ldots i_r j_1\ldots j_{N-r}} \varepsilon_{k_1\ldots k_r j_1\ldots j_{N-r}} A_{i_1\ldots i_r} A_{k_1\ldots k_r}$$

$$= \sum_{i_1<\ldots<i_r} A_{i_1\ldots i_r} A_{i_1\ldots i_r} = A * A$$

which is the desired result.

Having proved that D_r is an isomorphism from $\hat{T}_r(V)$ to $\hat{T}_{N-r}(V)$ relative to the inner product $*$, we can now use the condition (9.136) and (9.139) to compute the inverse of D_r. Indeed, from the skew-symmetry of the wedge product we have

$$A \wedge Z = (-1)^{r(N-r)} Z \wedge Z \tag{9.143}$$

Hence (9.136) yields

$$(D_r A) * Z = (-1)^{r(N-r)} A * (D_{N-r} Z) \tag{9.144}$$

for all $A \in \hat{T}_r(V)$ and $Z \in \hat{T}_{N-r}(V)$. On the other hand. (9.139) yields

$$(D_r A) * (Z) = A * (D_r^{-1} Z) \tag{9.145}$$

when we chose $D_r B$=Z. Comparing (9.144) with (9.145), we obtain

$$D_r^{-1} = (-1)^{r(N-r)} D_{N-r} \tag{9.146}$$

which is the desired result.

We notice that the equations (9.142) and (9.83) are very similar to each other, the only difference being that (9.83) applies to all bases while (9.142) is restricted to right-handed orthonormal bases only. Since the quantities given by (9.83) transform according to the rule of an axial tensor density, while the dual $D_r A$ is a tensor in $\hat{T}_{N-r}(V)$, the formula (9.142) is no longer valid if the basis is not right-handed and orthogonal, If $\left\{e^i\right\}$ is an arbitrary basis, then E is given by (9.125). In this case if we represent A, Z, and $D_r A$ again by their strict components relative to $\{e^i\}$, then from (9.83) and (9.122) the condition (9.136) can be written as

$$\sum_{\substack{i_1<\ldots<i_r \\ j_1<\ldots<j_{N-r}}} A_{i_1\ldots i_r} Z_{j_1\ldots j_{N-r}} \varepsilon^{i_1\ldots i_r j_1\ldots j_{N-r}} e^*$$

$$= \sum_{\substack{k_1<\ldots<k_{N-r} \\ j_1<\ldots<j_{N-r}}} \det\begin{bmatrix} e^{k_1 j_1} & \cdot & \cdot & \cdot & e^{k_1 j_{N-r}} \\ \cdot & & & & \cdot \\ \cdot & & & & \cdot \\ \cdot & & & & \cdot \\ e^{k_{N-r} j_1} & \cdot & \cdot & \cdot & e^{k_{N-r} j_{N-r}} \end{bmatrix} (D_r A)_{k_1\ldots k_{N-r}} Z_{j_1\ldots j_{N-r}} \quad (9.147)$$

Since Z is arbitrary, (9.147) implies

$$\sum_{i_1<\ldots<i_r} A_{i_1\ldots i_r} \varepsilon^{i_1\ldots i_r j_1\ldots j_{N-r}} e^*$$

$$= \sum_{k_1<\ldots<k_{N-r}} \det\begin{bmatrix} e^{k_1 j_1} & \cdot & \cdot & \cdot & e^{k_1 j_{N-r}} \\ \cdot & & & & \cdot \\ \cdot & & & & \cdot \\ \cdot & & & & \cdot \\ e^{k_{N-r} j_1} & \cdot & \cdot & \cdot & e^{k_{N-r} j_{N-r}} \end{bmatrix} (D_r A)_{k_1\ldots k_{N-r}} \quad (9.148)$$

We can solve (9.148) for the components of $D_r A$ in the following way. We multiply (9.148) by

$$\det\begin{bmatrix} e_{j_1 l_1} & \cdot & \cdot & \cdot & e_{j_1 l_{N-r}} \\ \cdot & & & & \cdot \\ \cdot & & & & \cdot \\ \cdot & & & & \cdot \\ e_{j_{N-r} l_1} & \cdot & \cdot & \cdot & e_{j_{N-r} l_{N-r}} \end{bmatrix}$$

and sum on $j_1,\ldots,j_{N-r}$ in increasing order, then we obtain

$$(D_r A)_{l_1 \ldots l_{N-r}}$$

$$= \sum_{\substack{i_1 < \ldots < i_r \\ j_1 < \ldots < j_{N-r}}} A_{i_1 \ldots i_r} e * \varepsilon^{i_1 \ldots i_r j_1 \ldots j_{N-r}} \det \begin{bmatrix} e_{j_1 l_1} & \cdot & \cdot & \cdot & e_{j_1 l_{N-r}} \\ \cdot & & & & \cdot \\ \cdot & & & & \cdot \\ \cdot & & & & \cdot \\ e_{j_{N-r} l_1} & \cdot & \cdot & \cdot & e_{j_{N-r} l_{N-r}} \end{bmatrix} \tag{9.149}$$

which is the desired result. In deriving (9.149) we have used the identity (6.42) for the matrix $[e^{ij}]$ and its inverse $[e_{ij}]$ in order to obtain the formula

$$\sum_{j_1 < \ldots < j_{N-r}} \det \begin{bmatrix} e^{k_1 j_1} & \cdot & \cdot & \cdot & e^{k_1 j_{N-r}} \\ \cdot & & & & \cdot \\ \cdot & & & & \cdot \\ \cdot & & & & \cdot \\ e^{k_{N-r} j_1} & \cdot & \cdot & \cdot & e^{k_{N-r} j_{N-r}} \end{bmatrix} \det \begin{bmatrix} e_{j_1 l_1} & \cdot & \cdot & \cdot & e_{j_1 l_{N-r}} \\ \cdot & & & & \cdot \\ \cdot & & & & \cdot \\ \cdot & & & & \cdot \\ e_{j_{N-r} l_1} & \cdot & \cdot & \cdot & e_{j_{N-r} l_{N-r}} \end{bmatrix} = \delta^{k_1 \ldots k_{N-r}}_{l_1 \ldots l_{N-r}} \tag{9.150}$$

Equation (9.149) follows, since we have

$$\sum_{k_1 < \ldots < k_{N-r}} \delta^{k_1 \ldots k_{N-r}}_{l_1 \ldots l_{N-r}} (D_r A)_{k_1 \ldots k_{N-r}} = (D_r A)_{l_1 \ldots l_{N-r}} \tag{9.151}$$

Notice that (9.142) is a special case of (9.149) when the basis is right-handed and orthonormal, since in this case $e^* = 1, e_{ij} = \delta_{ij}$, and, from (6.21),

$$\det \begin{bmatrix} \delta_{j_1 l_1} & \cdot & \cdot & \cdot & \delta_{j_1 l_{N-r}} \\ \cdot & & & & \cdot \\ \cdot & & & & \cdot \\ \cdot & & & & \cdot \\ \delta_{j_{N-r} l_1} & \cdot & \cdot & \cdot & \delta_{j_{N-r} l_{N-r}} \end{bmatrix} = \delta^{j_1 \ldots j_r}_{l_1 \ldots l_r}$$

Then as in (9.151) we have

$$\sum_{j_1 < \ldots < j_{N-r}} \varepsilon^{i_1 \ldots i_r j_1 \ldots j_{N-r}} \delta^{j_1 \ldots j_r}_{l_1 \ldots l_r} \varepsilon^{i_1 \ldots i_r l_1 \ldots l_{N-r}}$$

And thus (9.149) reduces to (9.142).

The duality operator can be used to define the *cross product* or the *vector product* for an oriented three-dimensional space. If we take $N = 3$, then D_2 is an isomorphism from $\hat{T}_2(V)$ to $\hat{T}_1(V)$

$$D_2 : \hat{T}_2(V) \to \hat{T}_1(V) = V \tag{9.152}$$

so for any u and $v \in V, D_2(u \wedge v)$ is a vector in V. We put

$$u \times v \equiv D_2(u \wedge v) \tag{9.153}$$

called the *cross product* of u with v.

We now prove that this definition is consistent with the classical definition of a cross product. This fact is more or less obvious. From (9.136), we have

$$E*(u \wedge v \wedge w) = u \cdot (v \times w) \qquad \text{for all } w \in V \tag{9.154}$$

where we have replaced the $*$ inner product on the right-hand side by the $\cdot$ inner product since $u \times v$ and w belong to V. Equation (9.154) shows that

$$(u \times v) \cdot u = (u \times v) \cdot v = 0 \tag{9.155$_1$}$$

so that $u \times v$ is orthogonal to v. That equation shows also that $u \times v = 0$ if and only if u, v are linearly dependent. Further, if u, v are linearly independent, and if n is the unit normal of u and v such that $\{u, v, n\}$ from a right-handed basis, then

$$(u \times v) \cdot n > 0 \tag{9.155$_2$}$$

which means that $u \times v$ is pointing in the same direction as n. Finally, from (9.127) we obtain

$$\begin{aligned}
[E*(u \wedge v \wedge w)][E*(u \wedge v \wedge w)] &= (u \times v) \cdot (u \times v) \\
&= \det \begin{bmatrix} u \cdot u & u \cdot v & 0 \\ v \cdot u & v \cdot v & 0 \\ 0 & 0 & 1 \end{bmatrix} = \|u\|^2 \|v\|^2 - (u \cdot v)^2 \\
&= \|u\|^2 \|v\|^2 (1 - \cos^2 \theta) = \|u\|^2 \|v\|^2 \sin^2 \theta
\end{aligned} \tag{9.156}$$

where θ is the angle between u and v. Hence we have shown that $u \times v$ is in the direction of n and has the magnitude $\|u\|\|v\|\sin\theta$, so that the definition of $u \times v$, based on (9.154), is consistent with the classical definition of the cross product.

From (9.153), the usual properties of the cross product are obvious; e.g., $u \times v$ is bilinear and skew-symmetric in u and v. From (9.142), relative to a right-handed orthogonal basis $\{i_1, i_2, i_3\}$ the components of $u \times v$ are given by

$$(u\times v)_k = \sum_{i<j} \varepsilon_{ijk}(u_i v_j - u_j v_i) = \varepsilon_{ijk} u_i v_j \tag{9.157}$$

where the summations on i, j in the last term are unrestricted. Thus we have

$$\begin{aligned}(u \times v)_1 &= u_2 v_3 - u_3 v_2\\ (u \times v)_2 &= u_3 v_1 - u_1 v_3\\ (u \times v)_3 &= u_1 v_2 - u_2 v_1\end{aligned} \tag{9.158}$$

On the other hand, if we use an arbitrary basis $\{e^i\}$, then from (9.149) we have

$$(u\times v)_t = \sum_{i<j} (u_i v_j - u_j v_i) e^* \varepsilon^{ijk} e_{kj} = e^* e_{kl} \varepsilon^{ijk} u_i v_j \tag{9.159}$$

where e^* is given by (9.125), namely

$$e^* = \varepsilon(\det[e^{ij}])^{1/2}$$

Exercises

1. If $D \in \hat{T}_N(V)$, what is the value of $D_N D$?
2. If $\{e^i\}$ is a basis of V, determine the strict components of the dual $D_r(e^{i_1} \wedge \cdots \wedge e^{i_r})$. *Hint:* The strict components of

   $e^{i_1} \wedge \cdots \wedge e^{i_r}$ are $\left\{\delta^{i_1 \ldots i_r}_{j_1 \ldots j_r}, j_1 < \cdots < j_r\right\}$ since as in (9.151) we have

$$\sum_{j_1 < \ldots < j_r} \delta^{i_1 \ldots i_r}_{j_1 \ldots j_r} e^{j_1} \wedge \cdots \wedge e^{j_r} = e^{i_1} \wedge \cdots \wedge e^{i_r} \tag{9.160}$$

3. If A is an endomorphism of a three-dimensional oriented inner product space V, show that

$$Au \cdot (Av \times Aw) = (\det A) u \cdot (v \times w) \tag{9.161}$$

and if A is invertible, show that

$$Av \times Aw = (\det A)(A^{-1})^T v \times w \tag{9.162}$$

Transformation to Contravariant Representation

So far we have used the covariant representations of skew-symmetric tensors only. We could, of course, develop the results of exterior algebra using the contravariant representations or even the midst representations of skew-symmetric tensors, since we have a fixed rule of transformation among the various representations based on the inner product.

In this section, we shall demonstrate the transformation from the covariant representation to the contravariant representation.

Recall that in general if A is a tensor of order r, then relative to any reciprocal bases $\{e_i\}$ and $\{e^i\}$ the contravariant components $A^{i_1 \dots i_r}$ of A are related to the covariant components $A_{i_1 \dots i_r}$ of A by

$$A^{i_1 \dots i_r} = e^{i_1 j_2} \cdots e^{i_r j_r} A_{j_1 \cdots j_r} \tag{9.163}$$

This transformation rule is valid for all rth order tensors, including skew-symmetric ones. However, for skew-symmetric tensors it is convenient to use the strict components.

Their transformation rule no longer has the simple form (9.163), since the summations on the repeated indices $j_1 \cdots j_r$ on the right-hand side of (9.163) are unrestricted. We shall now derive the transformation rule between the contravariant and the covariant strict components of a skew-symmetric tensor.

Recall that the strict components of a skew-symmetric tensor are simply the ordinary components restricted to an increasing set of indices, as shown in (9.72).

In order to obtain an equivalent form of (9.163) using the strict components of A only, we must replace the right-hand side of that equation by a restricted summation. For this purpose we use the identity

$$A_{j_1 \dots j_r} = A_{[j_1 \dots j_r]} = \frac{1}{r!} \delta^{k_1 \dots k_r}_{j_1 \dots j_r} A_{k_1 \dots k_r} \tag{9.164}$$

where, by assumption, A is skew-symmetric. Substituting (9.164) into (9.163), we obtain

$$A^{i_1 \dots i_r} = \frac{1}{r!} \delta^{k_1 \dots k_r}_{j_1 \dots j_r} e^{i_j j_1} \dots e^{i_r j_r} A_{k_1 \dots k_r} \tag{9.165}$$

Now, since the coefficient of $A_{k_1 \dots k_r}$ is skew-symmetric, we can restrict the summations on $k_1 \dots k_r$ to the increasing order by removing the factor $1/r!$, that is

$$A^{i_1 \dots i_r} = \sum_{k_1 < \dots < k_r} \delta^{k_1 \dots k_r}_{j_1 \dots j_r} e^{i_1 j_1} \cdots e^{i_r j_r} A_{k_1 \dots k_r} \tag{9.166}$$

This is the desired transformation rule from the covariant strict components to the contravariant ones. Clearly, the inverse of (9.166) is

$$A_{i_1 \dots i_r} = \sum_{k_1 < \dots < k_r} \delta^{j_1 \dots j_r}_{k_1 \dots k_r} e_{i_1 j_1} \cdots e_{i_r j_1} A^{k_1 \dots k_r} \tag{9.167}$$

which is the transformation rule from the contravariant strict components to the covariant ones.

From (6.37) we see that (9.166) is equivalent to

$$A^{i_1 \dots i_r} = \sum_{k_1 < \dots < k_r} e^{i_1 \dots i_r, k_1 \dots k_r} A_{k_1 \dots k_r}$$

$$= \sum_{k_1 < \dots < k_r} \det \begin{bmatrix} e^{i_1 k_1} & \cdot & \cdot & \cdot & e^{i_1 k_r} \\ \cdot & & & & \cdot \\ \cdot & & & & \cdot \\ \cdot & & & & \cdot \\ e^{i_r k_1} & \cdot & \cdot & \cdot & e^{i_r k_r} \end{bmatrix} A_{k_1 \dots k_r} \tag{9.168}$$

while (9.167) is equivalent to

$$A_{i_1 \ldots i_r} = \sum_{k_1 < \ldots < k_r} e_{i_1 \ldots i_r k_1 \ldots k_r} A^{k_1 \ldots k_r}$$

$$= \sum_{k_1 < \ldots < k_r} \det \begin{bmatrix} e_{i_1 k_1} & \cdot & \cdot & \cdot & e_{i_1 k_r} \\ \cdot & & & & \cdot \\ \cdot & & & & \cdot \\ \cdot & & & & \cdot \\ e_{i_r k_1} & \cdot & \cdot & \cdot & e_{i_r k_r} \end{bmatrix} A^{k_1 \ldots k_r} \tag{9.169}$$

In particular, when $r = N$, we have

$$A^{1 \ldots N} = \det[e^{ij}] A_{1 \ldots N}, \qquad A_{1 \ldots N} = \det[e_{ij}] A^{1 \ldots N} \tag{9.170}$$

From the transformation rules (9.168) and (9.169), or directly from the skew-symmetry of the wedge product, we see that the product basis of any reciprocal bases $\{e_i\}$ and $\{e^i\}$ obeys the following transformation rules:

$$e^{i_1} \wedge \cdots \wedge e^{i_r} = \sum_{k_1 < \ldots < k_r} \det \begin{bmatrix} e^{i_1 k_1} & \cdot & \cdot & \cdot & e^{i_1 k_r} \\ \cdot & & & & \cdot \\ \cdot & & & & \cdot \\ \cdot & & & & \cdot \\ e^{i_r k_1} & \cdot & \cdot & \cdot & e^{i_r k_r} \end{bmatrix} e_{k_1} \wedge \cdots e_{k_r} \tag{9.171}$$

and

$$e_{i_1} \wedge \cdots \wedge e_{i_r} = \sum_{k_1 < \ldots < k_r} \det \begin{bmatrix} e_{i_1 k_1} & \cdot & \cdot & \cdot & e_{i_1 k_r} \\ \cdot & & & & \cdot \\ \cdot & & & & \cdot \\ \cdot & & & & \cdot \\ e_{i_r k_1} & \cdot & \cdot & \cdot & e_{i_r k_r} \end{bmatrix} e^{k_1} \wedge \cdots e^{k_r} \tag{9.172}$$

In deriving (9.171) and (9.172) we have used the fact that the strict covariant components of $e^{i_1} \wedge \cdots \wedge e^{i_r}$ are

$$\{\delta^{i_1 \ldots i_r}_{j_1 \ldots j_r}, j_1 < \cdots < j_r\}$$

and, likewise, the strict contravariant components of $e_{i_1} \wedge \cdots \wedge e_{i_r}$ are

$$\{\delta^{j_1\ldots j_r}_{i_1\ldots i_r}, j_1 < \cdots < j_r\}$$

as shown by (9.160). If we apply (9.171) and (9.172) to the product bases $e^1 \wedge \cdots \wedge e^N$ and $e_1 \wedge \cdots \wedge e_N$, we get

$$\begin{aligned} e^1 \wedge \cdots \wedge e^N &= \det[e^{ij}] e_1 \wedge \cdots \wedge e_N \\ e_1 \wedge \cdots \wedge e_N &= \det[e_{ij}] e^1 \wedge \cdots \wedge e^N \end{aligned} \tag{9.173}$$

The product bases

$$\{e^{i_1} \wedge \cdots \wedge e^{i_r}, i_1 < \cdots < i_r\} \text{ and } \{e_{i_1} \wedge \cdots \wedge e_{i_r}, i_1 < \cdots < i_r\}$$

are reciprocal bases with respect to the inner product $*$, since from (9.92) we have

$$(e^{i_1} \wedge \cdots \wedge e^{i_r}) * (e_{j_1} \wedge \cdots \wedge e_{j_r}) = \det \begin{bmatrix} \delta^{i_1}_{j_1} & \cdot & \cdot & \cdot & \delta^{i_1}_{j_r} \\ \cdot & & & & \cdot \\ \cdot & & & & \cdot \\ \cdot & & & & \cdot \\ \delta^{i_r}_{j_1} & \cdot & \cdot & \cdot & \delta^{i_r}_{j_r} \end{bmatrix} = \delta^{i_1\ldots i_r}_{j_1\ldots j_r} \tag{9.174}$$

In particular, when $r = N$ we have

$$\left(e^1 \wedge \ldots \wedge e^N\right) * \left(e_1 \wedge \ldots \wedge e_N\right) = 1 \tag{9.175}$$

From(9.174), we can compute the $*$ inner product of any two rth order skew-symmetric tensors

$$A * B = \sum_{k_1 < \cdots < k_r} A^{k_1\ldots k_r} B_{k_1\ldots k_r} = \sum_{k_1 < \cdots < k_r} A_{k_1\ldots k_r} B^{k_1\ldots k_r} \tag{9.176}$$

These formulas are equivalent to the formula (9.93), which is based on the covariant strict components of A and B.

For an oriented space we have defined the density e* of a basis $\left\{e^i\right\}$ to be the components of $e^1 \wedge \ldots \wedge e^N$ relative to the positive unit density E, namely

$$e^1 \wedge \ldots \wedge e^N = \text{e}^*\text{E} \tag{9.177}$$

as shown by (9.122). Clearly we can define a similar component for the basis $\{e_i\}$, namely

$$e_1 \wedge \cdots \wedge e_N = eE \tag{9.178}$$

Further, from (9.175) the components e and e″ are related by

$$ee^* = 1 \tag{9.179}$$

In view of this relation, we call e the volume of $\{e_i\}$. Then $\{e_i\}$ is positively oriented or right-handed if its volume e is positive; otherwise, $\{e_i\}$ is negatively oriented or left-handed. As before, a unimodular basis $\{e_i\}$ is defined by the condition that the absolute volume $|e|$ is equal to unity.

We can compute the absolute volume by

$$e^2 = \det\left[e_{ij}\right] \tag{9.180}$$

or equivalently

$$|e| = \left(\det\left[e_{ij}\right]\right)^{1/2} \tag{9.181}$$

The proof is the same as that of (9.123) and (9.124). Substituting (9.181) into (9.179), we have also

$$e_1 \wedge \cdots \wedge e_N = \varepsilon\left(\det\left[e_{ij}\right]\right)^{1/2} E \tag{9.182}$$

where ε is + if $\{e_i\}$ is positively oriented and it is – if $\{e_i\}$ is negatively oriented.

Using the contravariant components, we can simplify the formulas somewhat; e.g., from (9.168) we can rewrite (9.177) as

$$(D_r A)^{j_1 \cdots j_{N-r}} = \sum_{i_1 < \cdots < i_r} A_{i_1 \ldots i_r} e^* \varepsilon^{i_1 \ldots i_r j_1 \cdots j_{N-r}} \tag{9.183}$$

which is equivalent to

$$(D_r A)_{j_1 \cdots j_{N-r}} = \sum_{i_1 < \cdots < i_r} A^{i_1 \ldots i_r} e\varepsilon_{i_1 \ldots i_r j_1 \cdots j_{N-r}} \tag{9.184}$$

Similarly, (9.159) can be rewritten as

$$(u \times v)^k = e^* \varepsilon^{ijk} u_i v_j \tag{9.185}$$

which is equivalent to

$$(u \times v)_k = e\varepsilon_{ijk} u^i v^j \tag{9.186}$$

Exercises

1. Prove the formula (9.184).
2. Use the result of Exercise 9.136 or the transformation rules (9.183)and (9.184) and show that

$$D_r\left(e^{i_1} \wedge \cdots \wedge e^{i_r}\right) = \sum_{j_1 < \cdots < j_r} e^* \varepsilon^{i_1 \ldots i_r j_1 \ldots j_{N-r}} e_{j_1} \wedge \cdots \wedge e_{j_{N-r}} \tag{9.187}$$

which is equivalent to

$$D_r\left(e_{i_1} \wedge \cdots \wedge e_{i_r}\right) = \sum_{j_1 < \cdots < j_r} e^* \varepsilon_{i_1 \ldots i_r j_1 \ldots j_{N-r}} e^{j_1} \wedge \cdots \wedge e^{j_{N-r}} \tag{9.188}$$

3. Show that has the representations

$$E = \frac{1}{e} \varepsilon^{i_1 \ldots i_N} e_{i_1} \otimes \cdots \otimes e_{i_N} = e\varepsilon_{j_1 \ldots j_N} e^{j_1} \otimes \cdots \otimes e^{j_N} \tag{9.189}$$

where

$$\frac{1}{e} \varepsilon^{i_1 \ldots i_N} = e\varepsilon_{j_1 \ldots j_N} e^{i_1 j_1} \ldots e^{i_N j_N} \tag{9.190}$$

and

$$e = \varepsilon\left(\det\left[e_{ij}\right]\right)^{1/2} \tag{9.191}$$

Bibliography

Adamek, J.: *Locally Presentable and Accessible Categories*, Cambridge University Press, Cambridge, 1994.

Alfsen, E. M. and Schultz, F. W.: *Geometry of State Spaces of Operator Algebras*, Boston-Basel, Berlin, 2003.

Awodey, S. and Butz, C.: *Topological Completeness for Higher Order Logic*, Journal of Symbolic Logic, 2000.

Awodey, S.: *Structure in Mathematics and Logic: A Categorical Perspective*, Philosophia Mathematica, 1996.

Baez, J. and Dolan, J.: *From Finite Sets to Feynman Diagrams*, Beyond, Berlin, 2001.

Baez, J.: *An Introduction to n-Categories*, Springer Verlag, Berlin, 1997.

Baianu, I. C.: *Categories, Functors and Quantum Algebraic Computations, in P. Suppes*, Congress Logic Mathematics Philosophy of Science, Bucharest, 1971.

Barr, M. and Wells, C.: *Category Theory for Computing Science*, CRM. Montreal, 1999.

————: *Toposes, Triples and Theories*, Springer Verlag, New York, 1985.

Bell, J. L.: *A Primer of Infinitesimal Analysis*, Cambridge University Press, Cambridge, 1998.

————: *The Continuum in Smooth Infinitesimal Analysis*, Kluwer, Dordrecht, 2001.

Bell, J. L.: *Toposes and Local Set Theories: An Introduction*, Oxford University Press, Oxford, 1988.

——————: *Category Theory and the Foundations of Mathematics*, British Journal for the Philosophy of Science, 1981.

Blass, A.: *The Interaction Between Category Theory and Set Theory*, Providence, AMS, 1984.

Borceux, F.: *Handbook of Categorical Algebra*, Cambridge University Press, Cambridge, 1994.

Bucur, I. and Deleanu A.: *Introduction to the Theory of Categories and Functors*, J. Wiley and Sons, London, 1968.

Cartan, H. and Eilenberg, S.: *Homological Algebra*, Princeton University Press, Pinceton, 1956.

Cartan, H.: *Homological Algebra*, Princeton University Press, Princeton, 1956.

Chevalley, C.: *The Theory of Lie Groups*, Princeton University Press, New Jersey, 1946.

Cohen, P. M.: *Universal Algebra*, Harper and Row, Tokyo, 1965.

Connes, A.: *Noncommutative Geometry*, Academic Press, New York, 1994.

Crole, R. L.: *Categories for Types*, Cambridge University Press, Cambridge, 1994.

Dirac, P. A.: *The Principles of Quantum Mechanics*, Clarendon Press, Oxford, 1930.

Dolan, J. and Baez, J.: *From Finite Sets to Feynman Diagrams*, Springer, Berlin, 2001.

Dolan, J.: *Higher-Dimensional Algebra III n-Categories and the Algebra of Opetopes*, Advances in Mathematics, 1998.

Eilenberg, S. and Cartan, H.: *Homological Algebra*, Princeton University Press, Princeton, 1956.

Eilenberg, S. and Steenrod, N.: *Foundations of Algebraic Topology*, Princeton University Press, Princeton, 1952.

Freyd, P.: *Functor Theory (Dissertation)*, Princeton University, Princeton, New Jersey.1960.

——————: *Abelian Categories: An Introduction to the Theory of Functors*, Harper & Row, New York, 1964.

——————: *Categories, Allegories*, North Holland, Amsterdam, 1990.

——————: *The Theories of Functors and Models*, Theories of Models, Amsterdam, 1965.

Gabriel, P. and Zisman, M.: *Category of Fractions and Homotopy Theory, Ergebnesse der Math*, Springer, Berlin, 1967.

Grothendieck, A.: *Séminaire de Géométrie Algébrique*, Springer Verlag, Berlin, 1997.

Hatcher, W.: *The Logical Foundations of Mathematics*, Pergamon Press, Oxford, 1982.

Healy, M.: *Category Theory Applied to Neural Modeling and Graphical Representations*, Computer Science Press, London, 2000.

Hyland, J.: *First Steps in Synthetic Domain Theory*, Springer, Berlin, 1991.

——————: *The Effective Topos*, North Holland, Amsterdam, 1982.

Isham, C. J.: *Spacetime and the Philosophical Challenges of Quantum Gravity*, Cambridge University Press, Cambridge, 2001.

Jacobs, B.: *Categorical Logic and Type Theory*, North Holland, Amsterdam, 1999.

Johnstone, P. T.: *How General is a Generalized Space?*, Cambridge University Press, Cambridge, 1985.

——————: *Sketches of an Elephant: A Topos Theory Compendium*, Oxford University Press, Oxford, 2002.

——————: *Stone Spaces*, Cambridge University Press, Cambridge, 1982.

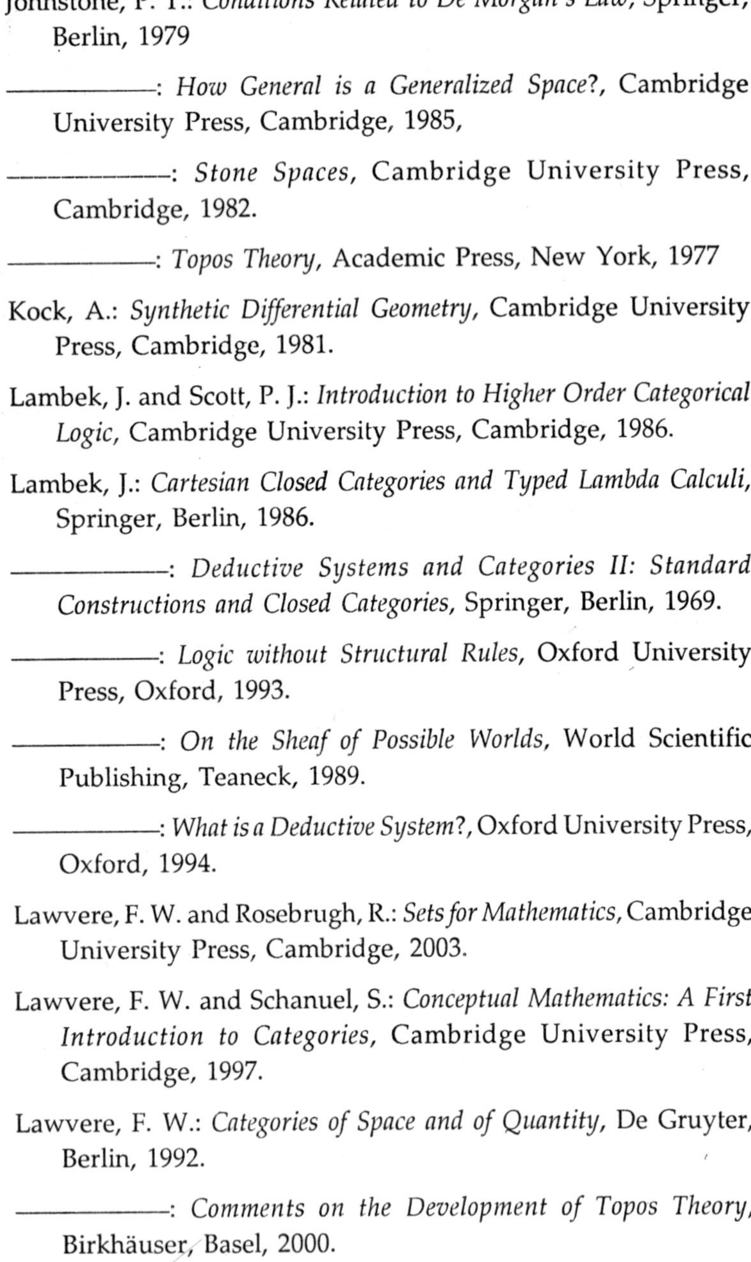

Johnstone, P. T.: *Conditions Related to De Morgan's Law*, Springer, Berlin, 1979

————: *How General is a Generalized Space?*, Cambridge University Press, Cambridge, 1985,

————: *Stone Spaces*, Cambridge University Press, Cambridge, 1982.

————: *Topos Theory*, Academic Press, New York, 1977

Kock, A.: *Synthetic Differential Geometry*, Cambridge University Press, Cambridge, 1981.

Lambek, J. and Scott, P. J.: *Introduction to Higher Order Categorical Logic*, Cambridge University Press, Cambridge, 1986.

Lambek, J.: *Cartesian Closed Categories and Typed Lambda Calculi*, Springer, Berlin, 1986.

————: *Deductive Systems and Categories II: Standard Constructions and Closed Categories*, Springer, Berlin, 1969.

————: *Logic without Structural Rules*, Oxford University Press, Oxford, 1993.

————: *On the Sheaf of Possible Worlds*, World Scientific Publishing, Teaneck, 1989.

————: *What is a Deductive System?*, Oxford University Press, Oxford, 1994.

Lawvere, F. W. and Rosebrugh, R.: *Sets for Mathematics*, Cambridge University Press, Cambridge, 2003.

Lawvere, F. W. and Schanuel, S.: *Conceptual Mathematics: A First Introduction to Categories*, Cambridge University Press, Cambridge, 1997.

Lawvere, F. W.: *Categories of Space and of Quantity*, De Gruyter, Berlin, 1992.

————: *Comments on the Development of Topos Theory*, Birkhäuser, Basel, 2000.

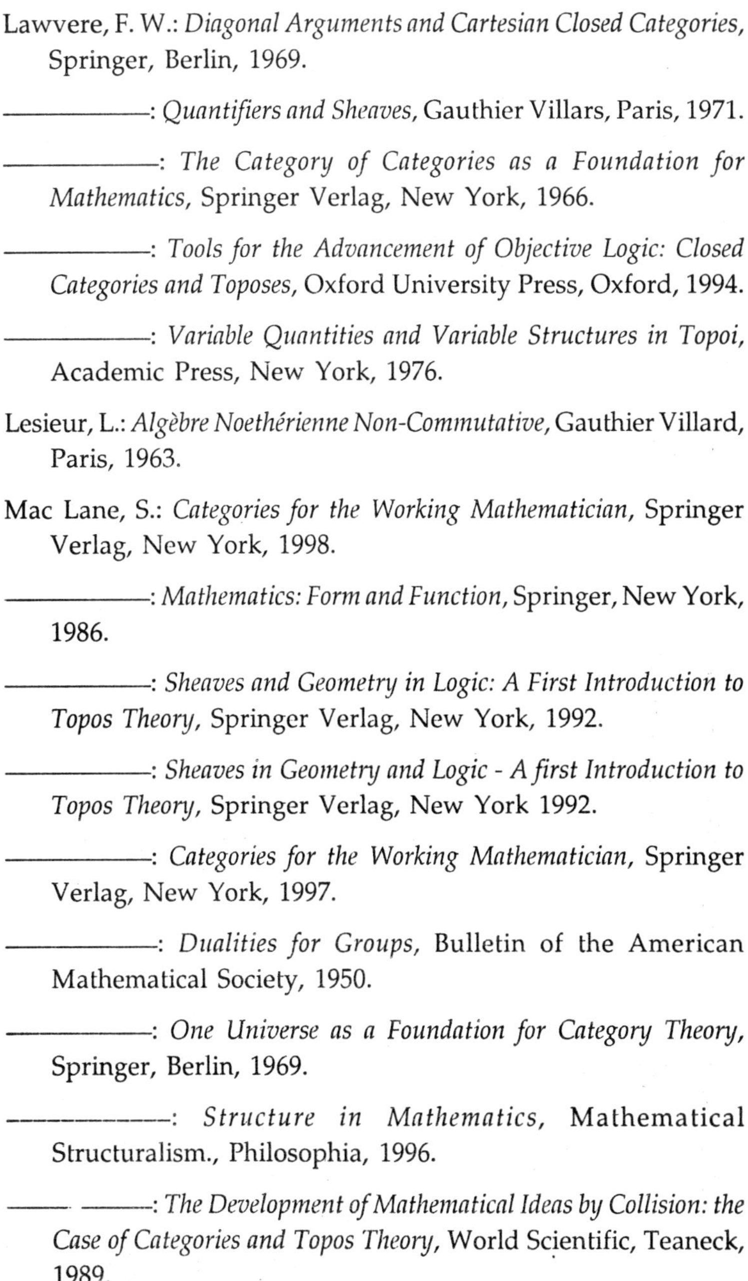

Lawvere, F. W.: *Diagonal Arguments and Cartesian Closed Categories,* Springer, Berlin, 1969.

————: *Quantifiers and Sheaves,* Gauthier Villars, Paris, 1971.

————: *The Category of Categories as a Foundation for Mathematics,* Springer Verlag, New York, 1966.

————: *Tools for the Advancement of Objective Logic: Closed Categories and Toposes,* Oxford University Press, Oxford, 1994.

————: *Variable Quantities and Variable Structures in Topoi,* Academic Press, New York, 1976.

Lesieur, L.: *Algèbre Noethérienne Non-Commutative,* Gauthier Villard, Paris, 1963.

Mac Lane, S.: *Categories for the Working Mathematician,* Springer Verlag, New York, 1998.

————: *Mathematics: Form and Function,* Springer, New York, 1986.

————: *Sheaves and Geometry in Logic: A First Introduction to Topos Theory,* Springer Verlag, New York, 1992.

————: *Sheaves in Geometry and Logic - A first Introduction to Topos Theory,* Springer Verlag, New York 1992.

————: *Categories for the Working Mathematician,* Springer Verlag, New York, 1997.

————: *Dualities for Groups,* Bulletin of the American Mathematical Society, 1950.

————: *One Universe as a Foundation for Category Theory,* Springer, Berlin, 1969.

————: *Structure in Mathematics,* Mathematical Structuralism., Philosophia, 1996.

————: *The Development of Mathematical Ideas by Collision: the Case of Categories and Topos Theory,* World Scientific, Teaneck, 1989.

MacNamara, J. and Reyes, G.: *The Logical Foundation of Cognition*, Oxford University Press, Oxford, 1994.

Majid, S.: *A Quantum Groups Primer*, Cambridge University Press, Cambridge, 2002.

————: *Foundations of Quantum Group Theory*, Cambridge University Press, Cambridge, 1995.

Makkai, M. and Reyes, G.: *First Order Categorical Logic*, Springer, New York, 1977.

Makkai, M.: *On Structuralism in Mathematics*, MIT Press, Cambridge, 1999.

————: *Towards a Categorical Foundation of Mathematics*, Springer, Berlin, 1998.

Marquis, J. P.: *Categories, Sets and the Nature of Mathematical Entities*, Springer, New York, 2006.

————: *Russell's Logicism and Categorical Logicisms*, University of Toronto Press, Toronto, 1993.

————: *Three Kinds of Universals in Mathematics?*, Mathematics and Science, Oxford, 2000.

————: *A Concise Course in Algebraic Topology*, The University of Chicago Press, Chicago, 1999.

Mc Larty, C.: *Category Theory in Real Time*, Philosophia Mathematica, 1994.

————: *Elementary Categories, Elementary Toposes*, Oxford University Press, Oxford, 1992.

————: *Exploring Categorical Structuralism*, Philosophia Mathematica, 2004.

Mitchell, B.: *Theory of Categories*, Academic Press, London, 1965.

Moerdijk, I.: *Algebraic Set Theory*, Cambridge University Press, Cambridge, 1995.

Neumann, J.: *Mathematische Grundlagen der Quantenmechanik*, Springer, Berlin, 1932.

Palme, R. M.: *Count Nouns, Mass Nouns, and their Transformations: a Unified Category-theoretic Semantics*, MIT Press, Cambridge, 1999.

Pareigis, B.: *Categories and Functors*, Academic Press, New York, 1970.

Paton, R. and Porter T.: *Categorical Language and Hierarchical Models for Cell Systems*, Springer Verlag, New York, 2004.

Pedicchio, M. C. and Tholen, W.: *Categorical Foundations*, Cambridge University Press, Cambridge, 2004.

Peirce, B.: *Basic Category Theory for Computer Scientists*, MIT Press, Cambridge, 1991.

Penrose, R.: *Shadows of the Mind*, Oxford University Press, Oxford, 1994.

Pitts, A. M.: *Categorical Logic*, Oxford Unversity Press, Oxford, 2000.

————: *Interpolation and Conceptual Completeness for Pretoposes via Category Theory*, Applied Mathematics, New York, 1987.

Plotkin, B.: *Algebra, Categories and Databases*, Handbook of Algebra, Amsterdam, 2000.

Popescu, N.: *Abelian Categories with Applications to Rings and Modules*, Academic Press, London, 1975.

Pribram, K. H.: *Brain and Perception: Holonomy and Structure in Figural Processing*, Lawrence Erlbaum Association, Hillsdale, 1991.

Prigogine, I.: *From Being to Becoming: Time and Complexity in the Physical Sciences*, W. H. Freeman and Co., San Francisco, 1980.

Reyes, G.: *First Order Categorical Logic*, Springer, New York, 1977.

—— ——: *From Sheaves to Logic*, Daigneault, Providence, 1974.

————: *Models for Smooth Infinitesimal Analysis*, Springer Verlag, New York, 1991.

Roberts, J. E.: *More Lectures on Algebraic Quantum Field Theory in A. Connes*, Springer, New York, 2004.

Robinson, P. L. and Plymen, R. J.: *Spinors in Hilbert Space*, Cambridge University Press, Cambridge, 1994

Rosen, R.: *Anticipatory Systems*, Pergamon Press, New York, 1985.

Scedrov, A.: *Freyd's Model for the Independence of the Axiom of Choice*, Providence, AMS, 1989.

Schrödinger E.: *Mind and Matter* in *'What is Life?'*, Cambridge University Press, Cambridge, 1967.

Scott, P. J.: *Some Aspects of Categories in Computer Science*, Amsterdam, North Holland, 2000.

————: *Category Theory for Linear Logicians*, Cambridge University Press, Cambridge, 2004.

Smolin, L.: *Three Roads to Quantum Gravity*, Basic Books, New York, 2001.

Spanier, E. H.: *Algebraic Topology*, McGraw Hill, New York, 1966.

Spencer, Brown G.: *Laws of Form*, George Allen and Unwin, London, 1969.

Stapp, H.: *Mind, Matter and Quantum Mechanics*, Springer Verlag, New York, 1993.

Steenrod, N.: *Foundations of Algebraic Topology*, Princeton University Press, Princeton, 1952.

Stewart, I. and Golubitsky, M.: *Fearful Symmetry: Is God a Geometer?*, Blackwell, Oxford, 1993.

Tattersall, I. and J. Schwartz: *Extinct Humans*, Westview Press, Oxford, 2000.

Taylor, P.: *Practical Foundations of Mathematics*, Cambridge University Press, Cambridge, 1999.

Thompson, W. D.: *On Growth and Form*, Dover Publications, New York, 1994.

Tierney, M.: *Sheaf Theory and the Continuum Hypothesis*, Springer Lecture Notes in Mathematics, New York, 1972.

Unruh, W. G.: *Black Holes, Dumb Holes, and Entropy*, Cambridge University Press, Cambridge 2001.

Van der Hoeven, G. and Moerdijk, I.: *Constructing Choice Sequences for Lawless Sequences of Neighbourhood Functions*, Springer, Berlin, 1984.

Vanbremeersch, J. P.: *Hierarchical Evolutive Systems: a Mathematical Model for Complex Systems*, Bulletin of Mathematical Biology, 1987.

Varela, F. J. and Maturana, H. R.: *Autopoiesis and Cognition: The Realization of the Living*, Reidel Publication, Dordrecht, 1980.

Wallace, R.: *Consciousness: A Mathematical Treatment of the Global Neuronal Workspace*, Springer, Berlin, 2005.

Weinberg, S.: *The Quantum Theory of Fields*, Cambridge University Press, Cambridge, 1995.

Wells, C.: *Toposes, Triples and Theories*, Springer Verlag, New York, 1985.

Wess, J. and Bagger, J.: *Supersymmetry and Supergravity*, Princeton University Press, New Jerseu, 1983.

Wiener, N.: *The Human Use of Human Beings: Cybernetics and Society*, Free Association Books, London, 1989.

Wood, R. J.: *Categorical Foundations*, Cambridge University Press, Cambridge, 2004.

Zolfaghari, H.: *Topos-theoretic Approaches to Modality*, Springer, Berlin, 1991.

Zurek, W.: *Quantum Theory and Measurement*, Princeton University Press, New Jerseu, 1983.

Index

D

E

F

G

H

I

K

L

M

N

O

P

Q

R

S

T

U

V

Z

□□□